Perri Pivovar

About the Author

ANNIE CHOI was born and raised in the greater Los Angeles area. She received her B.A. from the University of California, Berkeley, and her M.F.A. in writing from Columbia University. She has worked as a tour guide, elevator operator, assistant medical photographer, sign language teacher, and a science textbook editor. She lives in New York City. Her Web site is www.annietown.com

Happy Birthday or Whatever

Happy Birthday or Whatever

Track Suits, Kim Chee, and Other Family Disasters

ANNIE CHOI

HARPER

NEW YORK • LONDON • TORONTO • SYDNEY

HARPER

HarperCollins books may be purchased for educational, business, or sales promotional use. For information please write: Special Markets Department, Harper-Collins Publishers, 10 East 53rd Street, New York, NY 10022.

FIRST EDITION

Designed by Justin Dodd

Library of Congress Cataloging-in-Publication Data
Choi, Annie.
   Happy birthday or whatever : a memoir / Annie Choi.—1st ed.
      p. cm.
   ISBN: 978-0-06-113222-3
   ISBN-10: 0-06-113222-5
   1. Choi, Annie, 1976—Humor. 2. Korean Americans—Biography. 3. Immigrants—United States—Biography. 4. New York (N.Y.)—Biography. 5. California—Biography. I. Title.

CT275.C5725A3 2007
974.7'1004957092—dc22   2006043658

07 08 09 10 11 ❖/RRD 10 9 8 7 6 5 4 3 2 1

FOR MY PARENTS

contents

Happy Birthday or Whatever   1

Animals   15

Spelling B+   29

Crimes of Fashion   45

Stroke Order   63

Period Piece   77

Holy Crap   91

The Best Diet   109

Vegetarian Enough   129

The Devil Moisturizes   141

Fool Who Play Cool   161

Rules of Engagement   191

New Year's Games   209

Acknowledgments   241

HaPPY
BIrtHDaY
or
wHatever

HaPPY BIrtHDaY
or wHatever

I was going to have the best birthday ever. It would start with a parade—a dizzying spectacle of floats, prancing palominos, and the country's loudest marching bands. There would be troupes of mimes and contortionists, foul-mouthed drag queens, and a man juggling little girls on fire. Monkeys dressed in powder blue tuxedos would throw candy and tiny bottles of whiskey to the hordes of my fans lined up along Sixth Avenue. A dozen Michael Jackson impersonators, from his pre-op "Rock with You" days to his current noseless incarnation, would handle the sixty-foot helium balloon version of me. As the Grand Marshal, I would ride on the back of an elephant and wave as streamers, confetti, and twenty-dollar bills cascaded over me. After the procession, my friends and I would drink all the liquor in Manhattan, break tequila bottles over our heads, and pick fights with the Hell's Angels. The next morning we would crawl into work at the crack of noon, nursing hangovers and picking glass out of last night's clothes, and proclaim that the only birthday that could've been more historic was Jesus' bar mitzvah.

The morning of my twenty-seventh birthday, I received several e-mails from friends sheepishly bowing out of dinner, bar-hopping, and whatever mischief the night might bring. "No problem," I replied, "more liquor for the rest of us." Later, two more friends cancelled: "But maybe we'll make it—call us later tonight." No matter, I thought, the rest of us can still level every bar in the city. Then another friend explained he was "just too tired." I called him a geriatric and crossed him off my list. Seeing the members of my posse dwindle, I called my remaining friends to confirm our night of debauchery. One got a last-minute ticket to something that wouldn't be as exciting as my birthday—Madonna and her fake British accent in concert—and the other didn't return my calls. A half-hour before party time, other friends decided that meeting deadlines outweighed meeting Jose Cuervo. What would have been a highly intemperate party with twelve of my closest friends ended up being a quiet group of four (myself included) dining at a restaurant where tables were set with too many forks. We split a bottle of wine and ate outside. It was humid. Our waiter scraped breadcrumbs off the tablecloth with a little metal scoop, and the butter, which was sculpted into a tiny rose, sweated in the August evening heat. I turned twenty-seven with no monkeys or transvestites or celebrity impersonators. And no phone calls from my parents.

The next morning I woke up and checked my messages. Perhaps my parents called in the middle of the night; they live in Los Angeles and the three-hour time difference worked in their favor. Nothing. I was surprised; my parents use the phone as a 3,000-mile-long umbilical cord, and most of the time I want to strangle myself with it. My mother calls just to inform me that rice is on sale at Ralph's, but it's still cheaper to buy it at a Korean grocery store and how much is rice in New York and why do Americans eat Uncle Ben's, when he's not even Asian (one of the many things about Americans that still confuse my mother even though she and my father immigrated in 1971). On the one day that my parents were supposed to call, they didn't. Even my brother, the guy who used to wrestle me to the ground and fart in my face, remembered. Mike sent me a characteristically terse e-mail: "Happy birthday or whatever." My brother, in addition to being a master wordsmith, keeps untraditional hours. He spends his days sleeping and his nights processing loans for a major bank. But even he managed to send his little sister a birthday greeting.

I checked the missed-calls list on my cell phone. Nothing. What kind of parents forget their child's birthday? Bad ones. Unloving ones. Ones that don't deserve the World's Greatest Kid. (I have the mug to prove it; I stole it from my brother.) My birthday should be easy to remember—August 25th, the day before their wedding anniversary.

I toyed with the idea of not calling my parents on their anniversary and playing out a childish drama, but I realized that my parents never made a big deal about their anniversary. They've always appreciated my phone call and greetings, but I don't think they expected it. When I was growing up, August 26th was just another day. But on August 25th, I was the center of everyone's universe, and the Anniverse included a stuffed animal, my favorite meal (spaghetti or tofu stew), and an ice cream cake from Baskin-Robbins (Jamoca Almond Fudge). I flipped open my phone. At the very least, I could make my parents feel guilty. That could be fun. I dialed my mother first.

"Hello?"

"Hi, it's me. How are you?"

"Who this?"

I thought that evolution and genetics allowed parents to easily identify the voices of their offspring. This is why wolves can identify their pups by their whines and barks from miles away. My mother apparently opted out of that gene. Instead she got the one that suddenly made her forget the date she squeezed out a squirming eight-pound ball of flesh after spending nine months with an indomitable case of hemorrhoids, which she has always made a point of mentioning to me.

"This is your daughter."

"Anne?"

"Is there another?"

"Hi, Anne! You sound funny."

"Maybe a little older? More mature?"

"No. How you are?"

"I'm good. I'm calling to say happy anniversary."

Silence. I heard scratchy Korean AM radio playing in her car— a commercial for the new, faster Hyundai Sonata. The last time my mother was this quiet it was 1982, and she was heavily sedated after three root canals. She shuffled around the house slowly and groaned, just like a zombie, only with more drool.

"Mom? You there? Happy anniversary."

"No, today not Mommy anniversary."

She has always called it "Mommy anniversary." Evidently it is the day she got married without my father.

"Uh, yes today is your anniversary."

"No, Mommy anniversary in September."

"No, your birthday is in September. Dad's birthday is in September, too. But your anniversary is today."

"I don't think so, Anne. You wrong."

"Trust me, today is your anniversary. Do you know how I know it's your anniversary?"

"How?"

"It's the day after my birthday. Which was yesterday, the twenty-fifth ."

Silence again. Ah hah! This was the part where she cried and begged for forgiveness and shipped me an ice cream cake. With two flavors.

"Nooo, Anne, you birthday in September."

"What? Mom, my birthday is not in September."

"No, you wrong, Anne, you wrong."

"You're kidding, right? Don't you think I'd know when my own birthday is? Here, let me read you my driver's license."

"No Anne, I tell you, you birthday in September. I remember. How I forget?"

I rolled my eyes. My mother can be one smooth-talking and wily lady, but there was no way she could convince me and the Department of Motor Vehicles that my birthday was in September. I fumbled for my wallet.

"Let's see, it says here D-O-B. Do you know what that stands for? Now I don't have a dictionary handy, but I believe it stands for Date of Birth. So, according to the State of New York, I was born eight/twenty-five/seventy-six. That means my birthday was August twenty-fifth. Yesterday."

"Yesterday?"

"Yes. Yesterday. You totally forgot."

"Are you sure?"

"MOM, you forgot. Just admit that you're wrong and you forgot."
Silence again. I had her. She was trapped. Now I could taunt
her at will—point my finger, make faces, make demands: Bake me
some cookies. Take me to Disneyland. Buy me a pony.

"Oh . . . Anne . . . Mommy feel . . . bad. I such . . . terrible
mommy." She fell quiet again and breathed deeply. I felt a burning
twinge in my stomach. I tried to ignore it but it crept up inside me. I
hoped that it was acid reflux from last night's sweaty butter flower,
but I knew it was guilt. She didn't mean to forget. It was just a
birthday. No big deal. Plenty of parents all over the world have forgotten their children's birthdays. John Hughes even made a movie
about it. It's not like my mother didn't love me. She's done worse.

"Look, I'm sorry. You know I don't care, I just like to give you
a hard time." I could hear the soft whir of traffic and the radio.
Classical music.

"Oh good! OK. I drive now. Mommy go bye!"

Click.

"Ah, shit!" I snapped my cell phone closed. How the hell did
she do it? She wronged me, yet I was the one who'd apologized.
She sensed my weakness and moved in for the kill. She is a crafty
one, my mother. I simmered. I fumed. I mashed the buttons on my
phone and dialed my father's office. Someone should apologize to
me, damnit. The line was busy. My father doesn't believe in call
waiting. I suspected that my mother hung up on me and called him
immediately to caution him about their slighted daughter. She's
always one step ahead of me. Five minutes later, I got through to
my father.

"Hi, Dad."

"Oh, Anne, happy birthday! How are you! How's New York?"

I raised an eyebrow. He seemed too cheerful. I don't trust cheerful people. They are either elementary-school teachers or axe-murderers or parents who have forgotten something important.

"My birthday was yesterday."

"I know."

"So why didn't you call?"

"I very busy at lab. I mean to call you."

"You were too busy to call your own daughter on her birthday?" I don't think my father realized that remembering my birthday and not finding the time to call was actually worse than forgetting altogether. He was lying. He has always been bad at lying. When I was in third grade, my father cut a deal that if I earned straight A's on my report card, he'd quit smoking. I brought home straight A's, stuck my report card to the refrigerator, and demanded that he empty his carton into the garbage can. A week later I smelled cigarettes on his clothes. He denied it, even when I pointed to the familiar rectangular lump in his shirt pocket. I was crushed.

"You know, Mom didn't call me either."

"Really? Are you sure you not miss message?" I rolled my eyes. He's not as slick as my mother.

"Oh come on, you both forgot my birthday. Just admit it."

"I tell you Annie, I not forget. You birthday is day before my anniversary."

"Did Mom tell you that?"

My father laughed. He was caught. He knew it. "Maybe. Maybe not."

"Well, happy anniversary. Whatever."

"And happy birthday to you."

"It's too late to say happy birthday." I pouted. I was six years old again. Except when I was six my parents remembered my birthday.

"You sad, Annie?"

I could sense my father smiling and sulked even more, the way kids get angrier when parents laugh at the cute way they get angry. "No, no I'm not sad. Is there anything else you want to tell me?"

Sorry. My apologies. Excuse me. Forgive me. Please don't be angry with me, I am a terrible person and I don't know how I'm going to live with myself with all this guilt and shame and how can I make it up to you? May I interest you in a bucket of cash?

"No, that's it. I have to work now."

I hung up. My phone conversations with my father are never more than two minutes long because he's too busy putting out fires at work—literally. He's a metallurgical chemist with very little regard for safety. His hands are blistered from open chemicals and flames and part of his yellowed thumbnail is missing. Every shirt he owns is stained and full of holes from acid splatters. His face masks, goggles, aprons, and gloves remain in their original 1980s boxes, with the faded pictures of men in hard hats who are much too excited about safety. When I was eight, he gave me a glass mercury thermometer, which I liked to chew on and wave like a wand. Eventually he remembered that mercury causes mental retardation. When I was sixteen, I bought nail polish remover and Bactine, and my father made me return them, "I have better stuff at lab." The next day he brought home acetone and hydrogen peroxide in little plastic bottles labeled with their chemical formulas, C_3H_6O and H_2O_2. Incidentally, these are the only two formulas I ever got right on my chemistry tests. I guess in the excitement of labeling the bottles, my father forgot to dilute them. The laboratory-strength acetone was so strong it actually took off the nail polish plus the top layer of my fingernails, leaving ten little chalky splotches behind. I didn't dare try the peroxide. I showed my mother the remains of my fingertips and she shook her head: "You daddy such NERD!"

About two weeks after my parents' anniversary is my father's birthday. For a long time, the family celebrated his birthday at El Torito, Sizzler, or if he was feeling fancy, Black Angus. They were all conveniently located within a block of each other, near the mall. We would pile into the car with my father behind the wheel and he'd make a last-minute decision. My brother always lobbied for El Torito and its Conquistador platter, which was a combinación magnífico of everything stuffed, battered, and fried on the menu. After going vegetarian in high school, I became particularly fond of Sizzler, with its generous salad bar and sundae station. I called my mother to remind her and show off the fact that I was a good person—a better person, perhaps even the best. I remembered birthdays. I was a thoughtful daughter and not at all a little brat wanting to be appreciated. No, not at all.

"Hello?"

"Hi Mom, it's me, your daughter, Annie Choi. Do you remember me? We've met a few times."

"Oh you make Mommy laugh. How I get such funny daughter? I leave Costco now. I buy butter."

"You went all the way to Costco just to get that? Why didn't you just go to Ralph's?" I paused, trying to think of a clever segue. "Do you know what day it is?" Hmm, not so clever.

"I don't know. Mommy very late for haircut."

She was en route to her Koreatown hair salon, which has frenetic Korean pop music, slick black furniture, and mirrors on every surface. My mother's stylist is a wispy young Korean woman with thick fake eyelashes that look a lot like pubic hair. My mother's salon is supposedly geared toward young Korean hipsters, but it is usually occupied by scheming Korean ladies who sit under heat lamps and discuss their perfect, expensively educated, doctor/lawyer children. It is here my mother believes she will find me a loving husband who will buy me a German luxury sedan.

"Mom, today is September ninth. Do you know what that means?"

"You birthday?"

"WHAT? My birthday was—"

"—August twenty-sixth, Mommy know. I kidding, Annie. Joke. Oh, you so serious."

"MOM, August twenty-sixth is your anniversary. I'm the twenty-fifth. You are totally hopeless."

"Annie, I say August twenty-fifth."

"That's a lie—you're lying, you liar. You said August twenty-sixth. I heard you."

She laughed, and I heard her golf clubs jostle violently in the back of her S.U.V. My mother refuses to pay attention to lane lines in parking lots. She likes to cut across the entire thing, honking and swerving around people and shopping carts that get in her way.

"No, no, I know what I say, Anne. You so silly!"

"Mom, today is September ninth."

"Yes, I know."

"Think, what happens on September ninth?"

"I told you, Mommy get haircut."

"No. Something important."

"I have golf lesson."

"Uh, not quite. Here is a hint. It starts with a 'D' and ends in 'ad's birthday.'"

I imagine an MRI of her brain, with a huge black blob taking over everything, except for the sections that control golf and hair care. Surely medical professionals will be puzzled by her gross memory loss, but impressed by her powerful backswing and her lustrous bob.

"I not understand. What you mean?"

"Today is Dad's birthday. Remember birthdays? They come once a year for each person in our family. His happens to be today.

We wouldn't want to forget his birthday, would we? That'd be horrible. So horrible. I can't even imagine anything worse."

She was unresponsive.

"OK, Anne, don't forget call you daddy and say happy birthday."

I groaned. I called my father at his lab. "Hi, Dad. Happy birthday!"

"Annie, thank you."

"See, I remembered. Look how well you raised me, look how thoughtful I am. I hope you're proud. How old are you?"

"Too old—sixty-one."

"Hey, you can retire soon."

"I'll never retire. Someone has to pay for Mommy golf." He laughed because he knew it was true.

"Are you doing anything special? El Torito? A parade?"

"What? Who get a parade for birthday? Only Santa Claus get a parade. I want nothing. I want to do nothing. I go home early, watch TV, fall asleep."

"Sounds exciting."

"I'm sixty-two. I'm getting old."

"I thought you said you were sixty-one"

"Oh, Annie, sixty-one or sixty-two. Makes no difference to me."

One week after my father's birthday is my mother's. We'd celebrate her birth the only way we knew how—by going to El Torito, Sizzler, or Black Angus. She never had a preference, as long as she didn't have to cook anything. On September 16th I called my mother. This afternoon she was at home, which surprised me. I assumed she'd be out with her church friends, getting their hair done (again) or playing golf or returning shoes at Nordstrom's.

"Happy birthday, Mom! You're fifty-five!"

"More young than fifty-six, but more old than fifty-four."

"Well, that's generally how it works. What are you up to today?"

"I got car wash."

"Whoa, take it easy there. Did you get drunk?"

"Anne, is that what you do in New York? Go party? Get drunk? You think fun? Do drug?"

"No, no, Mom, I'm kidding."

"I get so worry. You know Mrs. Lee son in trouble because he buy drug. Very bad. I read in newspaper Korean kid do drug and kill his mommy and daddy. So crazy. And one Korean girl she go to Princeton but she die because someone drink and hit with car. Big car."

My mother is a devoted reader of the *Korea Times,* which is the largest Korean-language newspaper published in America and has regional editions that mostly focus on local stories. A few weeks ago she mentioned that the front page lauded a Torrance-area fourteen-year-old Korean who won the Academic Decathlon and went on tour with the L.A. Philharmonic as a solo violinist. "Why you not in newspaper?" she asked sullenly. I explained that the *Korea Times* would never write a story about a Korean girl from the San Fernando Valley who quit piano in eighth grade and left every light on in the house, costing her family hundreds of thousands of dollars in electricity.

"MOM, this is about you. Happy birthday. Why aren't you with your friends? What about Grandma?"

"She so old. Today I visit her and ask for one dollar for parking, she give me twenty."

"I don't see a problem there. Twenty is better than one. They teach you that in math."

"She so old, she can't hear."

"Well she wouldn't listen to you anyway. So what are you doing now?"

She sighed loudly. A little too loudly. "Mommy eat lunch alone. Rice with water . . . alone . . . by myself. Only little rice. Not even full bowl. All Mommy friend busy today."

I rolled my eyes. What a drama queen. Drama-rama. Drama-mine. Andramada. Draman noodles. "Why don't you go out to eat?"

"I use all my money for gas."

This was the gas she needed to fuel her car to sit in traffic to get to the salon, where she got a $110 haircut and blow-dry.

"What about Dad?"

"You daddy not even give me present."

"Well, did you give him a present for his birthday?"

"No."

"Why not?"

"Because the daddy supposed give gift to the mommy." She laughed, tickled by her own joke.

"No, I thought the mommy and the daddy were supposed to give gifts to the kids. No one gave me a gift."

"No, kid give gift to the mommy and daddy."

"Actually, we're supposed to give gifts to one another. That's what other families do."

"Maybe you right. But we not other family."

"Don't you think we should try?"

I remembered the birthday dinners when my mother made spaghetti in my honor. My father would eat it with chopsticks and *kim chee,* spicy pickled cabbage. After we polished off the ice cream cake, there was a mad dash to the bathrooms because we are all lactose intolerant.

ANIMALS

I was twenty-three when I met Arnold. It was at Dongdaemun, a congested flea market in Seoul where merchants peddle products of dubious quality and taste: fatigued radishes and cabbage, paper-thin underwear and socks, telephones shaped like high-heeled shoes, knock-off "Prado" handbags, couches upholstered with floral fabric that feels like sandpaper. Merchants bellow promises of low prices and large selections and search for the slightest sign of interest from passing shoppers. Having an unflappable poker face is necessary whenever I walk through Dongdaemun; with one wayward glance I can end up listening to a relentless sales pitch for a rooster-shaped clock that cock-a-doodle-doos every hour. I locked eyes with Arnold in one of the booths, where he sat comfortably between a scruffy dog wearing a red and white striped shirt and a parrot that looked a little constipated.

"Oh my God! Look at him!"

I stopped in mid-step and stared. I grinned broadly and let out a toe-curling squeal. I might as well have pulled out my wallet and waved my cash around. My mother, sensing trouble, tugged at my arm and tried to usher me past the booth, but it was too late. The *ajuma* quickly put down her bowl of rice and jumped off her stool. She wiped her hands on her apron and patted her pockets. One contained money and the other a wooden abacus.

"Anne, why you stop? Now ajuma come talk to us. You make Mommy so tire. *Ayoo* . . ."

Ayoo is what Koreans say when they experience chronic pain or annoyance. My mother says it with an exaggerated crescendo that instinctively causes me to scrunch up my shoulders and wince, just as I do when I see someone get kicked in the crotch: "*Ayoo*, Anne, why you hair so mess?" "*Ayoo*, why you not call Mommy?" "*Ayoo*, why you not go church? You make God mad." As I get older, I find myself ayooing my mother's ayoos.

The ajuma, with dollar, or rather, *won* signs in her eyes, readjusted the apron strings around her generous waist and waddled over to my mother and me. Seeing my mother's scowl, the merchant decided I was the easier target.

"Well, look at this beautiful young woman! You look like a smart girl—the smartest. You must know that I have the best selection, better than anyone else here—the largest. Which one would you like? I'll give you a good price—the best."

She followed my gaze and reached for the constipated parrot.

"No, no, I don't want the . . ."

I realized I didn't know how to say "parrot" in Korean. Or "constipated" for that matter. My listening skills have always been better than my verbal skills. I pointed to Arnold.

"Ah, that one? What a good choice, this is a fine animal—the best." The ajuma handed him to me.

My mother balked at my selection. "What it is?"

"It's a pig."

"No, I think maybe bear?"

"No way, it's pink. It's a pig. I think."

"*Ayoo*, Anne, it not pig. It ugly!"

My mother was right. It ugly. Everything about Arnold was lopsided. One arm was higher than the other. One ear was twisted and faced slightly backward. One leg was longer than the other, but both were too short for his lumpy torso. The seam running down his round belly was puckered and crooked. His head wobbled on a weak neck. His fur was soft but threadbare.

"Why you like this? I think maybe blind man make this!" She pointed to the feet, which were mismatched, misshapen blobs.

I looked at the tag attached to his crooked ear. His name written in Korean and I sounded out the letters slowly—*ah-nol-duh*. I hugged Arnold and tried to remember the last time I had bought a stuffed animal for myself.

"Oh, Anne, not this again. Why you do this? Do you want give Mommy ulcer?"

I was spoiled as a child. I was the youngest, the smallest, and therefore the cutest. Once at a party when I was six, my father's second oldest brother gave me a small shaggy dog with floppy ears. I was so pleased I told him that the dog was my favorite toy and he was my favorite uncle. He beamed proudly and swaggered around the apartment boasting about his title to the rest of the family.

"Did you hear? I'm Annie's favorite. I'm the best. She loves me the most. That means she loves you less!"

A few hours later, my father's younger brother brought me a polar bear so massive that I had difficulty wrapping my arms around it. Its fur was gleaming white and it had gigantic paws with black leather pads that squeaked when they were squeezed. My uncle had left the party just to go to a toy store and outshine his older brother. I jumped up and down and swung my new friend around by its squeaking paws, and I told my uncle he was now my favorite relative, unless someone else brought me something better. I looked around the room expecting another furry surprise. My mother tugged on my ponytail.

"Please, everyone," my mother announced in Korean, "don't buy our little Annie any more gifts. She's too spoiled and she's turning into a brat. And no one likes brats, isn't that right, Annie?"

I grimaced and buried my head into the polar bear's chest. My relatives laughed and an aunt winked at me. She handed me a box of Yan-Yans, which have pretzel sticks on one side and a cup of chocolate frosting on the other—my favorite treat.

Of all the toys I received, I responded most to stuffed animals. Barbies and Legos remained mostly untouched, and I shunned board games and sports equipment. As a five-year-old, I liked making up a history for the animals; some were saved from starvation in the desert, others were runaways from the circus. Most had neurotic tendencies—one animal didn't like the color orange, another was allergic to wool. They didn't even have to be animals. At an arcade, my cousin Woo-jay played one of those machines filled with cheap toys and a moveable claw. About one hour and ten bucks later he brought me a fuzzy striped ball that probably cost a dime to manufacture. It was blue and pink and I wrapped it in a blanket and carried it wherever I went. I dressed it up in a bonnet and put one of my sweaters on it. This embarrassed my mother, who feared that people would think she couldn't afford a real doll for her daughter. I named the ball Blink, a combination of "blue" and "pink." Weeks later, Woo-jay, who wanted to defend his title as my favorite cousin, brought me another fuzzy ball. This one was red and green, and I named it Reen. The two balls were long-lost friends and Reen was happy to be reunited after doing hard time in a glass booth at Chuck E. Cheese. When my brother tried to juggle them, I burst into tears.

"Stop it! Stop it! They don't like being thrown!"

"They're balls, stupid."

"They're afraid of heights."

"No they're not. They like it. See?" Mike tossed Blink and Reen high in the air and they hit the ceiling. Little pieces of cottage-cheese plaster came drifting down.

"STOP! You're going to make them throw up!"

As I collected more and more animals, I became anxious. They all sat on my bed but there were too many to put under the blankets with me, so I worried that they would get cold. I collected handkerchiefs and scarves and fabric samples—anything that could keep my animals warm. At first, each animal (or ball) had its own blanket. As the numbers increased, the animals shared: two to a blanket, then three, depending on size. I felt that four animals sharing a blanket would be cruel, so I moved on to the fancy cloth napkins I found in the china cabinet.

"Anne, this napkin for people. Not you doll!"

"But they're gonna get cold. They're gonna get sick."

"Why they get sick? They not live!"

My mother didn't seem to care that my animals were freezing, especially during August in Southern California, where temperatures could plummet to a paw-chilling 110 degrees. Shortly after, I discovered hand towels and my mother scolded me. So then I moved on to Kleenex (I liked the scented pink ones) and my t-shirts. Every morning while I was at kindergarten my mother would go into my bedroom and refold my shirts and stuff the tissues back in the box. Every afternoon when I returned, I got them out and tucked everyone in again.

Once during the middle of the night, my mother came and removed all the blankets on my animals. I woke up and immediately burst into tears. I was upset that the animals had gotten cold and felt like a horrible mother.

"Mom, you're mean. Say sorry to everyone!"

"Anne, I not say sorry. They only toy. Not real, you understand? Mommy think you crazy!"

"Why do you hate them? How come you don't love them?"

My mother gave up. She asked my relatives to pass on their hankies and my grandmother sewed little blankets. She even knitted an overalls and jacket set for a bear to keep him warm. He wore it, but still needed a blanket anyway.

On the bed, I lined up all my animals against the wall, and there was a specific order. Most were carefully organized by size. A larger animal sat to the left of a slightly smaller animal: The giant sheepdog with unruly hair, which I brushed until he went bald, sat to the left of the pudgy brown walrus my uncle bought me at Sea World, which sat next to a smaller hot pink lion. Animals that came into my possession together had to stay together. The two striped balls, for example, sat next to each other, and were later joined by a rainbow-colored one (named Rainball). When a new animal joined the group, I introduced him or her to everyone and then found the proper spot in the lineup. Often blankets had to be reorganized so everyone could get coverage.

At one point I realized some animals might get jealous if I spent too much time with one. I had to make sure I loved each one equally. I considered the spot closest to my head as the preferred spot; the animal there would receive the most attention and the most warmth. So each night, the animals rotated. The first animal in line would sleep with me and then move to the back of the line the next day. It was a simple, efficient system no adult, especially my mother, could understand.

Whenever my mother changed my sheets, the animals would wind up on the floor in a chaotic pile of matted tails and paws and ears. This sent me into a fit of tears and pounding fists, followed by a lecture from my mother, and a few more minutes of crying.

"You sick, Annie, very sick. This not good for you."

I would spend the rest of the afternoon organizing the bedroom zoo and apologizing to each animal, asking for forgiveness and admonishing myself for being a bad mother.

By the time I was seven years old, I had collected so many animals my bed was getting full. I reorganized them into several rows, which significantly decreased my sleeping space. I slept as close to the edge of the bed as possible so my animals would have more room. I rarely tossed and turned through the night for fear of squishing and suffocating them.

"Anne you sleep in small side of bed. You doll get too much room. Why you do this? Not good for you. Move doll to floor!"

"Animals do not belong on the floor. I don't sleep on the floor, so the animals shouldn't sleep on the floor."

"You know, in Korea everyone sleep on floor. You grandma sleep on floor."

"But they're Korean. These are animals. They're special."

When I was eight years old, my mother, brother, and I spent two weeks in Seoul with my aunt's family. My mother told me I could only take one small animal, one that could fit in my suitcase. It was horrible, this idea that I could only take one animal and leave over a dozen behind, and that this one "lucky" animal would have to travel in a suitcase where clearly there would not be enough air or light or even a buddy to keep it company. I spent an agonizing amount of time standing in front of the lineup, trying to make my choice, hoping the others wouldn't hate me the way I hated my mother for making me abandon everyone. Eventually I selected the koala bear in the Los Angeles Dodgers jersey. He had a baseball hat, with white hair exploding from his little tan ears and a tuft of white on his rear. I didn't particularly appreciate, or even understand, baseball, but I probably chose the koala because my father bought him for me—I wouldn't see my father for two weeks. The koala fit nicely between my clothes and a gigantic bag of banana chips. My mother always brought random American foodstuffs to Seoul, and I would've forgone the ten packs of beef jerky and the institutional-sized canister of Skippy peanut butter for another animal.

After two weeks in Seoul, it was time to leave my family and reunite with my loved ones. As we were packing to leave, my cousin Eun-hee, three years my senior, slyly mentioned she really liked my koala.

This is a problem in Korean families. When a relative says she "really likes" something that is a mere indulgence to you, perhaps a jacket or a pair of shoes or maybe a koala in an adorable baseball jersey, you are forced to hand it over. A few years ago I brought from London a silk scarf for my mother and a hat for my grandmother. After opening the gifts, my grandmother mentioned she "really liked" the scarf, and my mother suggested a trade. Then my grandmother mentioned she "really liked" the hat, too. So my grandmother walked away with both. Just like that. My mother was disappointed.

"Don't worry, Mom. You'll get both back next year when Grandma dies."

"ANNE! Sometime I think how I raise such devil?"

When Eun-hee mentioned she "really liked" my koala, I didn't yet know the way of the cunning Korean. My cousin told her mother, who told my mother, who told me that I would have to give up the koala. I did not understand this. Eun-hee was eleven years old. She did not need a koala; she needed a boyfriend. Koalas were for eight-year-olds like me. My mother had forced me to pick one animal and now I had to leave it in Korea, thousands of miles away from home, from his friends? To strand him in a country that didn't even speak his language? She was nuts. So I did what I could; I squirmed and cried wildly on the floor, my long hair streaked with snot and tears. And then I hid him deep in my suitcase.

As we were leaving for the airport, I saw a flash of white and brown and an unmistakable baseball hat in Eun-hee's arms. Somewhere, in my luggage, there was a koala-shaped hole. My mother had to drag my thrashing body into the car. She promised to buy me another koala in a Los Angeles Dodgers jersey. The same exact one, the kind you can only buy at Dodger Stadium, about a one-hour drive from our house, not including traffic. To this day, that woman still owes me a koala.

The koala kidnapping sparked a wave of animal abductions. I think my mother realized that the only way to force me to grow up was to give away what I loved the most. It was also a mental health issue. I spent a lot of time worrying about my animals and my mother figured if I didn't have any animals, then I wouldn't worry anymore. My parents often opened their house to friends they met at church or at my Korean Saturday school. Some were old college classmates who had recently moved to the United States. All these Koreans had kids who "really liked" my animals. I guess there is something irresistible about a white seal bundled up in a flowery handkerchief that once held the contents of my aunt's nose.

"Anne, Yoon like doll. You give to her."

"NOOOOOO!"

I loved that white seal. So did all the other animals. I couldn't part with it. Yoon couldn't possibly care for the white seal. She couldn't love it like me. Plus, she had dirty hands. I yelled, I pouted, I wept. I caused an ungracious stink right on the living room floor, in front of Yoon and all of my parents' guests. I acted like a brat, but try taking a cub away from its mother. She will not say, "Hey, that's cool, I have three more."

My conniption proved my mother's failure to raise a polite and selfless child, or at least one who refused to pretend to be polite and selfless in front of guests. On stage in the living room, in front of her friends and Yoon's mother, my mother had to demonstrate her child-rearing skills.

"Anne, give Yoon doll. Now."

"NO!"

"Don't make Mommy mad."

"NO. I DON'T WANT TO. IT'S MINE."

"Yoon have no doll. You have too many. You give her one."

"NO! NO! NO! NO! NO!"

"Anne, I get very angry. Upstair, *now.*"

Going upstairs meant a very serious lecture about The Right Thing. A few tissues and a smack on the rear later, I gave Yoon the seal.

"Please have my doll. I am sorry for being rude."

My mother ignored all my emotional explosions and contin-ued to give away my animals. If a little girl liked an animal, it was gone. If a young boy acted extra cute or extra smart, my animal would be the reward. My mother became everyone's favorite aun-tie. Children knew her for her generosity; their parents knew me for my beastly, territorial nature. My animals and I sensed doom whenever we smelled fancy Korean food, the kind made only for guests. As a safety measure, I began hiding the animals under the bed. My mother, seeing the conspicuous exodus from my bed, did not appreciate this.

"Anne why you not share? You so spoil!"

"I do too share, I give my animals away all the time!"

Though I did hide quite a few animals, the kids were smart. They had all heard about my animals available for take-out. They went on searches in my room; it was like burglars coming and over-turning furniture to find hidden booty. I watched anxiously as they compared my bear with my kitten, deciding which one they would tell my mother they "really liked." I tried to stop them, to divert their attention to something else, like the piano or the swing set in the backyard, things I could live without. It was useless; the kids zeroed in on the animals. Each party ended with me pouting in my room, my remaining animals and I tearfully comforting each other.

When my family and I visited other families, I would introduce myself to their animals and ask them about their history. Where did they come from? What was their favorite food? What grade were they in? It never occurred to me to "really like" someone else's animal when I went to a party with my parents. It seemed so evil to tear one away from its family, and I knew that even if I took one home, it would never truly be mine.

Sometimes after a party at our house, I'd go back to my room and sense things were horribly wrong—maybe an animal was in the wrong place or the line was shorter than usual—and it would take a moment to take animal attendance. When owners give away puppies or kittens, the mother keeps searching for them, restlessly looking under furniture and sniffing corners. She calls out to them and whines. Then slowly she spends less time searching. And then one day, she just stops. That's what I did. I looked everywhere, and then one day I finally gave up.

By the time I was in high school, I still slept with my animals on the bed, though there were noticeably fewer. Years of parties had reduced the animal population to near extinction. A few even moved to the floor to make room for my cat—a real cat, not a stuffed animal. I gave a few animals to cute toddlers that visited our house, and donated some to a toy drive. There were a few I couldn't part with. When I left for college, I brought along Sushi, a small red lobster who hid comfortably under a pillow. A few years ago, my parents packed up their house to move to a smaller one. My mother mentioned that she had thrown all my animals out.

"WHAT? You threw them ALL away?"

"You not here, what I do?"

"All of them? You didn't give them to anyone?"

"So dirty Annie! Who want old doll? Who? Not Mommy!"

My animals were too dingy to donate to charity. No kid wanted them; they didn't have electronic voices or walk on their own or have kung-fu grip. No one "really liked" them anymore. My animals were outdated. I felt guilty. I had failed my animals, failed to protect them from my mother. Now they were in the trash, probably very cold and disoriented, with rats nesting in their fur.

It wasn't until recently that friends and boyfriends started buying stuffed animals for me. Buford the dog and Milkshake the cow, both gifts from my best friend, sit on my bed. When a hippo came to me on my birthday five years ago, I named him Hep Zepi the Hippie-Hating Hip-Hop Hippo. Hep Zepi is the name of a hip-hop clothing store in Boston. The hippie-hating part originates from angst leftover from my stint at Berkeley. On Valentine's Day in 1998, a guy I was dating gave me Bümpé, an exotic red bull designed by a Swedish artist. The seams on his ample backside are weakening and I get nervous that too many hugs will cause his polyester/cotton fill to explode. And then there is Arnold, the pig I bought at Dongdaemun with my mother.

"I think maybe bear. No Annie, it dirty. Look it have stain on face!"

"Mom, come on, those are rosy cheeks."

She was not convinced. She cringed. "*Ayoo,* I think dirt. You not baby now. You not need bear."

The ajuma, wanting to please my mother and make a sale, showed me a knock-off Hello Kitty and a puce tiger that lacked the imperfections to give it personality. My mother was unimpressed. So was I.

"How much?" I asked.

"Nine thousand won."

Numbers in Korean are hard for me. I had to count off my fingers to translate her offer. Nine dollars. It was a bargain, but only a fool accepts the first offer at Dongdaemun. "No, no too much. How about . . ." *one, two, three, four, five . . .* "six thousand won."

"Annie, you can't buy. You waste you money."

The ajuma, sensing danger to a sale of a stuffed animal that clearly no one else would want, closed the deal. She scooped up Arnold and put him in a plastic bag. I pulled out my wallet and counted out the money carefully, making sure I didn't mistake the one thousand note for a ten thousand bill.

"Oh Annie, you too old. This bear so ugly."

"Mom, it's a PIG."

"Pig, bear, Mommy not care."

"Well then maybe you won't give this one away." I stuck my tongue out at my mother and she tried to grab it between her fingers.

When I packed my luggage to return home alone—my mother stayed in Korea longer—I decided to put Arnold in my carry-on. It just didn't seem right to stuff him in a suitcase. I walked around the shops at Kimpo Airport, buying duty-free liquor and sniffing cosmetics, while Arnold's head poked out of my backpack. Other customers smiled at me—a twenty-three-year-old with a slightly deformed, yet completely loveable pig/bear. I like to think he enjoyed the view.

SPELLING B+

My mother let out one long sigh that seemed to last forever. She pursed her lips into a thin pink line and shook her head slowly.

"*Ayoo,* Anne, what I tell you over and over?"

I wasn't sure if I was supposed to answer. I looked at her ruefully and sank down into my seat, hoping that the vinyl cushion on my kitchen chair would swallow me whole. Then I could live the rest of my days lounging in the soft cotton stuffing.

"You make Mommy so tire!" She rubbed her shoulders and the back of her neck. She turned her head from side to side and the loud cracking and crunching made me shudder. I imagined there were no bones inside her neck, only potato chips—salt and vinegar.

"Why you do this to Mommy and Daddy?"

I slumped farther down and scrunched my shoulders together, trying to make it easier for the cushion to swallow me. My mother knitted her dark eyebrows together. The wrinkles on her forehead were deep enough to hide quarters, maybe even dinner plates.

"How you get such bad grade?"

I studied my bowl of oxtail soup—clear broth with small chunks of beef and blobs of fat floating on top. I stirred slowly and watched a lonely onion snake around my spoon. A turnip bobbed up and down. My mother dropped her chopsticks to the table; they clattered and rolled toward me. She reviewed my spelling quiz again in disbelief. Her eyes shifted from left to right as she scrutinized every item. When she saw a big red X next to a misspelled word, she narrowed her eyes and grumbled.

"How you go through life with grade like this?" She waved my quiz in the air. The gigantic B+ my third-grade teacher had scrawled in atomic red ink irradiated our drab kitchen. "Anne, what I say all time? I feel like doll with string. You pull and it say same thing over and over. You have to study and do homework!"

I felt a familiar combination of shame and anger. Heat rose in the back of my throat and behind my eyes. "I *do* study, I *do* do my homework."

"Then you not study enough! You get B on test. If you study, you get A!"

"It's not a B, it's a B-*plus*. I studied and got a B-plus!"

"Anne, B is not A, only A is A."

"But it's a B-*plus*!"

"When you get B, no one happy. You have to study, get A, go Harvard, and be doctor. Or lawyer. Maybe dentist like Dr. Kim.

Then you take care of Mommy and Daddy when you older. Then everybody happy."

It was my duty to go to the most prestigious university in America: "Me and Daddy job to take care of you and Mike. You and Mike job to go Harvard." As far as my parents were concerned, there was no other university worth attending. My parents left Korea in search of better opportunities, and that included Harvard. Even Yale and Princeton were just imitations. I was ten years away from college, but my parents acted as if the admissions board would review all my elementary school quizzes and deny me their crimson glory because I spelled *decide* with an *s*.

"Anne, I tell you now, B-plus not A.

"It's close to an A."

"Close not good enough." She cleared a spot on the kitchen table and handed me a pencil and a pad of paper. "Write every word you get wrong ten time."

"Ten times? That's too many."

"ANNE! Don't make Mommy mad!"

My oxtail soup got cold as I copied each word. My mother scooted her chair beside me and watched over my shoulder. I felt the heat of her breath as she sighed.

"Why you handwrite so bad? You have to write slow and clear."

"I *am* going slow."

"No, you rush. You go too fast and make mistake. Just like when you play piano!"

I stewed in my seat. I hated piano. I hated spelling. I hated my teacher for giving me a B-plus. I hated my mother for thinking it wasn't good enough—that I wasn't good enough. And the only thing worse than oxtail soup was cold oxtail soup. I pressed down on my pencil hard.

"Anne, you do good in math and science. You teacher say you read good. But why you bad in spelling? I not understand."

"Mom, I'm not bad at spelling!"

Spelling was difficult for me, but I knew I wasn't a bad speller. I was a better-than-average, B-plus speller, which is very close to being a fairly outstanding, A-minus speller. I thought this was quite an accomplishment considering that English wasn't my first language.

"Anne, I worry. Maybe you fail spelling. From now on you copy every word ten time before test."

"But that's going to take forever! All night!"

"No complain! Can you spell *complain*?" She made me write it down. I wasn't sure if I spelled it correctly. Neither was she.

My mother instituted her new studying regimen. Every Thursday night she sat with me while I copied every word from my spelling list ten times. My third-grade teacher grouped the words by themes: "What's New at the Zoo?" (*lion, cage, feed*); "What's for Breakfast?" (*milk, banana, cereal*); or "Crazy for Clothes" (*dress, suit, sweater*). Occasionally, there were words she couldn't pronounce or understand.

"What this word?"

"It's *cul-de-sac*."

"What that?"

"It's when the street ends and there are just houses there and everyone can play in the street because there are no cars."

"Why this word have so many dash? Seem so silly!"

"I don't know. Why does *sign* have a g in it? See? Spelling is hard, right?"

"Anne, life is hard! But you right, spelling very hard. I feel lucky that you take test and not Mommy! I think I get B on test!"

"Maybe you'd get a C."

"No, no, I get B. No one get C! Who get C?"

"There are kids who get C's."

"I think they mommy and daddy be very disappoint. Can you spell *disappoint*?"

After I copied each word ten times, my mother gave me a "test for practice." She read every word from the list aloud and had me write them down. If I spelled a word incorrectly, I had to copy it another ten times. Then she tested me again.

"You practice until you spell everything right!"

"But my hand hurts from writing."

"Don't be baby! You not wear diaper no more. You have to practice so you get A."

The nights before my spelling tests, I started having nightmares where I showed up to class unprepared and received a B. I'd wake up relieved and terrified, with my bangs plastered to my clammy forehead. Then I'd reach for my word list and notepad to make sure I had studied. On the mornings of my quizzes, my mother drilled me while I brushed my hair, put on my clothes, ate breakfast, tied my shoelaces, and rode in the car to school. After I slammed the car door, my mother rolled down the window and yelled, "Get A and make Mommy and Daddy proud!" During the spelling tests my heart pounded so violently I could barely hear my teacher call out the words. My head swirled with consonants and vowels—e's switched places with i's, and g's could be noisy or silent. Does *hammer* have two m's or one? Is it *nickel* or *nickle*? I wrote the more difficult words in light, tiny print, hoping my teacher wouldn't be able to read my handwriting but would somehow mark them correct.

Despite my mental anguish, my mother's studying methods proved successful. I received perfect scores. Bright red A's with stars and smiley faces decorated the top of my tests.

"My daughter so good at spelling. I so proud. Can you spell *proud*?"

"P-r-o-u-d. That word's easy—e-a-s-y."

"Very good! How you get so smart?"

"S-m-a-r-t!"

My mother chuckled and tried to find other words to challenge me. She thought I could spell anything. So did I. I settled into a Thursday evening routine and set the autopilot for Cambridge. Around the same time I was mastering my word lists, my brother managed to do the impossible and brought home a B on a math test ("OH MY GOD HOW YOU GET B? I THINK I DIE!"). My mother, confident that I could study spelling without supervision, diverted all her resources to her troubled son. Getting a B in sixth grade, so late in his academic career, could mean a life of illiteracy at Yale, or even a state school— what my parents called a "no-name school" or a "junk school."

As the school year continued, the words on my spelling list got longer and harder. Words such as *horse* and *street* were replaced by *raccoon* and *healthy,* or in my case, *racoon* and *helthy.* I discovered that copying a word ten times wasn't enough to get an A. It wasn't even enough to get a B+. When I scored a B on a spelling test, I knew my mother would be disappointed. I could see the ivied gates of Harvard close and the neon "No Vacancy" sign switch on. I realized there was only one thing I could do—lie. I couldn't lose my mother's love and pride, not after all that work, all those hours of sitting at the kitchen table with a cramped hand. My carpool dropped me off and I entered nervously into the house. My mother was in the kitchen cleaning a whole fish.

"Hi, Anne, how school?" She looked up from the cutting board and smiled. I hated fish—the gummy, cloudy eyes of dead fish heads grossed me out. Sometimes my mother chased me around the kitchen with a fish head and taunted me.

"Fine."

"How you do on spelling test?"

"Good . . . I got an A." I forced a ridiculously wide grin. I knew I wasn't good at lying; I didn't practice enough. I shoved my hands in my pockets so their shaking wouldn't betray me. "I got nothing wrong!"

"One hundred percent again! My only daughter such genius! You spell better than dictionary!"

Suddenly it occurred to me that she might want to see my quiz. I panicked and thought of excuses: I lost it; I left it at school; I gave it to a friend; the teacher put it up on the bulletin board because it was perfect. I clenched my jaw to prevent my heart from leaping out of my mouth. Why didn't they teach lying at school? It was more useful than spelling. I searched my mother's face for suspicion. She smiled and returned to the cutting board.

"You want practice piano before or after dinner?" She hacked away the fish tail.

Relief washed over me and I nearly melted to the floor in a p-u-d-d-l-e. Getting perfect scores had become such a routine that my mother didn't bother to check my test. She trusted me.

"I'll practice before." I felt piano was my punishment for lying. Sometimes I thought that piano was my punishment for living. I had to practice an hour every day. I picked up my backpack and shuffled to my room, the weight of a B pressing on my shoulders. I took out the quiz from the bottom of my bag and cringed. I folded it into a tiny square and slipped it in a folder pocket. No, no she'd find it there. I buried it in my dresser drawer. No, no sometimes she likes to pick out my school clothes. I hid it in my stuffed polar bear's overalls. Then I hid the bear under my sheets. At night I felt Snowball judging me, his black, beady eyes gazing steadily at me as I tried to sleep. I promised myself and the bear that I'd do better on the next quiz.

The words continued getting harder and though I copied each ten times, I received a B− on the next spelling test. It was the lowest grade I had ever received in any subject. I thought I was flunking out of elementary school. I would wind up in the streets alone, living out of a shopping cart and holding a cup and a sign just like the men and women I saw in downtown L.A., except my sign would say "Will Werk for Food." My heart sank into my Reebok high-tops. My throat tightened and tears pooled in my eyes. My misspelled words dripped in blood-red ink: *thunnder, tornadoe, lightening.* I couldn't show my mother a B−. It would ruin her and destroy me. I hid the test beneath a large bottle of ant spray in a cabinet. I couldn't stand to have it in my bedroom. Besides, Snowball's overalls couldn't hold another test. I didn't think of throwing it in the trashcan outside. Under the insecticide, the test would remain out of sight and mind. And it did for just a few hours.

"OH, NO, NO, ANNIE! COME HERE NOW!"

In the kitchen, my mother was holding the ant spray in one hand and a spelling test in the other. On the kitchen table there was a half-eaten apple covered in pulsating black specks. The worst part was that it was *my* apple. I had forgotten to throw it away, even though I knew that our house had a problem with ants. I started weeping. Extra bile ate away at my eight-year-old stomach; it was just a matter of time before all my organs liquefied.

"OH MY GOD, ANNIE. WHY YOU HIDE TEST?"

"I don't know." My knees started shaking. I wanted to lie down and curl into a ball.

"Why you do this? Why hide bad grade?"

"I don't know. Because I knew you'd get mad." I stared at my feet through my tears. I was wearing white socks with purple ribbons.

"You think Mommy mad? You right, I very mad. You get C and lie—make me very angry!"

"It's a B-minus . . ."

"No, Annie, it not B-minus, it C! How you get C?"

"I don't know."

"You don't know? You get C and you don't know? No wonder you got C—you don't know! So bad, Annie, this very bad."

"I'm sorry, I'm so sorry."

"Oh Mommy can't believe. You get A before, now you fail. I so confuse. Why you lie to Mommy?"

My mother shook my quiz at me. The red letter swung back and forth, crying out deception and shame. I knew what came next. My heart quickened, and sweat and tears streamed down my hot, swollen face. I wondered where she was going to spank me, how hard, and with what. I stood there, my collar soaked with tears, waiting for her to grab a bamboo sushi mat or a rice spoon. She quietly spritzed my half-eaten apple with ant spray. Then she added a cruel twist.

"Go to kitchen. Give me something to spank."

All the muscles in my neck and arms went stiff. This was a new situation. Normally she would storm directly to the middle drawer under the stove and grab a spoon or a spatula. Or she would dash to the front door and find the long shoehorn. Or she would find my red Snoopy ruler. But this time, it was my choice. I couldn't make my legs move.

"Get it from kitchen. Now! Go!" She sprayed angrily. Ants scattered from the poison, wiggled around in circles, and then stopped dead.

I walked slowly toward the kitchen, my chest hiccupping from crying. I stalled and looked back at her, pleading with my puffy, pink eyes, trying to convince her that I had learned my lesson and didn't need a spanking and would never, ever lie again and would study even harder. Copy each word twenty times. A hundred even.

"ANNIE, why you wait? GO!"

I went into the kitchen and looked around. I knew that if I chose something soft, like a towel or basting brush, she'd get even angrier and find something herself—not a good idea. But, I didn't want it to be too painful. I stood in the kitchen, thinking about my options.

"Why you take so long? What you do?"

I opened a drawer. Barbecue tongs? No, not good. A cheese-grater? No, too dangerous. A rolling pin? Definitely not.

"Come back here now!"

I went with an old rice spoon. I was familiar with its sting. It was made out of dark wood and shaped like an oversized screw-driver. It used to have a long, thin handle, until it broke one day while my mother was spanking my brother and me after we had wrestled and pummeled each other with all the living room cush-ions. Years of mixing, scooping rice, and spanking children had taken its toll on the spoon, and the handle had snapped in half. Right after it broke, we were all silent. Even my mother was a little shocked; she seemed incredulous that she hit us with enough strength to split wood. Still, she smacked us each once more on the thighs, testing out the more aerodynamic and portable version. It suited her fine and fit better in the drawer, so she kept it around.

I walked back to the living room and nervously gave my mom the broken spoon. She seemed to approve of my choice.

"Annie, do you understand why you get spank?

"Yes . . ." I gulped and choked. My toes curled as I tried to prepare myself for the sting, but it wasn't the spanking that would hurt.

"Why?"

"B-b-because I hid the test."

"And why you hide test?"

"B-b-because I got a b-b-bad grade."

"You have to study harder. You have to get A. You have to promise never, ever lie to Mommy."

"I promise, I promise."

She hit my palms a few times and remembered it wasn't good for piano, so she moved on to my bottom. Doubled over her lap, I sobbed. My tears and drool collected in a small pool near her bare feet. I went up to my room, where my stuffed animals consoled me, and I slept.

The next day my mother brought home several glossy workbooks, each about an inch thick. She pushed them across the kitchen table. One of the covers pictured a classroom of gaptoothed children and a teacher. One kid reached his hand high in the air with a sense of urgency and excitement. He looked as if he needed to go to the bathroom.

"See, Annie, this for you. You get A in spelling now. Always. You do five page every week. And you do five page English and five page math."

"But I'm doing good in English and math—"

"ANNE!"

She called my extra worksheets "Mommy homework." Aside from the homework my teachers assigned in regular school and Korean Saturday school, I now had to complete fifteen pages of math problems, English exercises, and spelling worksheets. I noticed that the math and English workbooks were for the grade level above me, but I didn't protest. With the spanking still fresh in my mind, I silently completed my Mommy homework. I started receiving perfect scores on my spelling tests again—partly from the spelling workbooks, but mostly because my mother made me copy each word twenty times. She drilled me relentlessly—over every meal, in car rides to my piano lessons, during my baths, while I dried the dishes.

Near the end of the school year, my teacher held a class spelling bee. The top three finishers would compete in a school-wide contest later that month. Twenty-eight of my classmates and I stood in a line that wrapped around the room. Miss Jensen called out a word and if a student misspelled it, he or she had to take a seat. I watched as my classmates sat down one by one. I defeated them easily—*picnic, measure, astronaut*—they were all words I could spell backward. I was the last person standing, the best speller in class, and I enjoyed the glory for about two seconds. Then I realized that I'd be competing against older students in front of the entire school. There was no way I could win. The prize for winning the class spelling bee was to be humiliated in front of the whole school.

"Spelling bees are stupid. I hate spelling."

"Not stupid, Anne. It like game."

"It's a stupid game."

"You try. We try together. Maybe you win."

"I don't want to win."

"You have to try, Anne."

My teacher gave me a list of words I should study—words that appeared on the fourth-, fifth-, and sixth-grade spelling tests. My mother put the math and English workbooks on hold while we studied the words together: *manufacture, extraterrestrial, whimsical, bureau.* I had no idea what these words meant, but I learned to spell them all in a month.

On the morning of the school-wide spelling bee, I woke up with terrible stomach cramps. I writhed in the bathroom, wishing I could jump down the sink and end up in the Pacific Ocean, swimming with the dolphins. My mother knocked and slid my spelling list under the door. I whimpered. I never wanted to spell anything ever a-g-a-i-n; I hated words and what they had done to me. I started shaking. I was a quiet, shy kid in school and I preferred to die rather than stand on stage and spell *sincerely*.

"Anne, you come out now."

I opened the door so my mother could see my pale sickly face and maybe take pity on me and let me stay home from school for the rest of my life.

"Mom, I don't want to go."

"You have to go."

"I feel sick."

"No excuse. Go. Now. You be OK, I promise."

On stage there was a long table where three teachers sat. Each had a clipboard, and one had a dictionary the size of a mattress. There were almost forty contestants, ranging from third to sixth grade. We sat in chairs near the stage, while the rest of the school watched in boredom. Parents were not invited; I couldn't even imagine competing in front of my mother. When my name was called, I slowly approached the stage. At first, I hoped that I wouldn't fall. Then I hoped that I would fall, break every bone in my body, be whisked away in an ambulance, and never have to spell again.

"Spell *garage*."

I sighed with a combination of relief and dread. Relief because it was an easy word. Dread because I knew I could spell it and would have to spell another word. And possibly another one and another one.

"*Garage*. G-a-r-a-g-e. *Garage*." I whispered into the microphone.

"That is correct."

I shuffled toward my seat. This continued for nearly two hours. As the words got harder—*sculpture, capacity, ingenious*—students began misspelling words. I managed to spell everything correctly and I spelled my way to the final five students. In the final round, if a contestant misspelled a word, the next contestant had to spell it correctly and then spell a new word. I looked around at the last five students. I was the youngest; the others were fifth- and sixth-graders. I had beaten all the fourth-graders. I wore a Minnie Mouse sweatshirt while the other girls wore nail polish and lip gloss. I was scared of the boys; they looked like men. One had a mous-tache. I fidgeted in my seat until it was my turn again. I walked up to the microphone.

"Spell *lyre*."

Easy, I thought, I can do this one. If there was one word I could spell, it was this one. I relaxed just the tiniest bit.

"*Lyre*. L-i-a-r. *Lyre*,"

"I'm sorry, that is incorrect."

I was stunned. *Liar*. I knew how to spell it; they were lying to me. I walked back to my seat in a daze. How else can you spell *liar*? I fumed. They must've made a mistake. A lanky sixth-grade girl stepped to the microphone. She had long straight hair with feathered bangs and wore a jean skirt and ankle boots. She smiled confidently.

"*Lyre*. L-y-r-e. *Lyre*."

"That is correct."

What? That couldn't be right. I looked at the teacher with the oversized dictionary. She nodded and gave her approval. What is a lyre? I simmered in my seat. I came in fifth. After all that work, I lost.

My mother picked me up from school. I slammed the car door and jerked on my seat belt.

"Mom, what's a lyre?"

"Someone who not tell truth."

"How do you spell it?"

"L-i-a-r."

"Some girl spelled it l-y-r-e."

"Well then she spell wrong."

"No, she got it *right*. The teachers said so. What's a lyre?"

"I don't know. We look up at home."

We checked in our World Book Encyclopedia.

"Why didn't they just ask me to spell *harp*? That was no fair!"

"You supposed to ask for definition, Anne."

"It was a mean trick! I hate spelling. I should've won!"

My teacher approached me the next day, patted me on the back, and handed me a thick booklet. It was full of words—hundreds of words in alphabetical order. Some contained more vowels than any other word I had ever seen. Other words were packed with x's and q's, and my tongue couldn't curl enough to pronounce them. The booklet contained the words most frequently used in the National Spelling Bee. Because I came in fifth place, I was the second alternate contestant for the regional competition. In the case that one of the three finalists and the first alternate couldn't make regionals, I would go in their place. My mother and I knew that I didn't have a chance of going. I wasn't sure if I wanted to go anyway because I hated spelling. But just in case, I kept the booklet on my bedside table. Every night before I drifted off to sleep, I opened it to one of the pages in the back to read the most exotic and challenging words. My favorite was *ytterbium*. I wondered what it meant.

CRIMES OF FASHION

In one of my favorite photographs of my mother, she is standing in front of her college in Seoul with three friends. Her eyes are half closed and her mouth is hanging wide open with her tongue slightly lolling out. Her head is tilted a little backward, allowing a clear view of her nostrils, which, for some reason, look larger than normal. Her elbows jut out at awkward angles as her hands reach up toward her face, in effort to hide it before the photographer clicked the shutter button. It's quite an unflattering picture of my mother except for one thing: her clothes. She is wearing a striped A-line dress with a high collar and wide cuffs at the end of long, tight sleeves. The hemline floats well above her knee-high boots with stacked heels. Though the photograph is a little blurry and black and white, I can tell her outfit would have turned heads until they twisted right off their necks. Her friends, also caught in various states of graceless surprise, are wearing long, plain skirts and modest blouses. One young lady is wearing a boxy cardigan, most likely purchased in Mr. Rogers' neighborhood. Next to her friends, my mother looks modern—confident, independent, and sophisticated.

My mother always considered every moment in public as an opportunity to show off her sense of style, as if walking to the post office were like strutting down a catwalk. When she was young, friends and relatives called her a "Hollywood movie star" and asked for an autograph. Most Korean girls are raised to be demure and bookish, but my mother liked being the center of attention, and often demanded it. Clothing elevated her above the rest of the crowd and allowed her to distinguish herself as an individual, while still belonging to a special group—the glamorous and sophisticated elite. When she was older, my mother cringed at the mousy moms who wore sweats and flip-flops when they picked up kids from school: "They look like they just wake up!" My mother pulled up to my elementary school wearing a fitted blazer and pinstripe slacks, as if carpooling required business-professional attire. She felt that just because she was a stay-at-home mom, she didn't have to dress like one. She coordinated colorful silk blouses with sassy pleated skirts and Italian leather pumps. She wore sweaters in every style imaginable—V-neck, scoop neck, turtleneck, mock neck, cowl neck—and they came in mouth-watering colors that she described as lemon, lime, and raspberry. Even her socks were stylish— argyle, polka dots, stripes, checkers. One of my chores when I was young was to fold laundry and matching her socks was the only fun part.

The labels of my mother's clothes announced sophistication and a certain level of financial comfort, but she picked through department store sales and outlet malls to buy brand-name fashion at sensible prices. While I squirmed on the floor or rearranged the clothes by color, she browsed the racks for hours, sliding hundreds of dresses, shirts, and skirts from one side to the other. Every few minutes she would hold up a garment and tell me its designer, and then critique their latest line for the season.

"Ralph Lauren for fall look so silly, with tight pants and big boot. It look like clothes for riding horse! Who has horse? Only farmer has horse and he wear Osh Kosh!"

Going to the mall with my mother was like going to fashion school. At six years old, I learned that cashmere came from goats and was thinner and warmer than wool. Before I knew how to multiply, I knew that linen garments wrinkled easily at first, but after several washings they would become softer and less prone to creases. Throughout elementary school, I learned exactly when plaid was in and when it was out, and when it was in again, and I found out what kind of people wore certain designers. Chanel was for the "old lady with big hair and little dog." Calvin Klein was for the "girl who look hungry but not eat." Liz Claiborne was for "lady with big hip and big purse." Versace, she explained one afternoon at Nordstrom when I was nine, was for women who wore two diamond rings on each finger and fur during a Los Angeles winter.

"Veh-sahch-ee for lady who get many plastic surgery."

"What's plastic surgery?"

"When you old you get plastic surgery to make young face. Young face, with no wrinkle. You know, wrinkle, like what you give Mommy!"

She laughed and showed me a Versace blouse. It was silk with red, bloated sleeves and a frenetic print of yellow lassos, blue flags, and white life preservers. It had oversized golden buttons in the shape of anchors.

"Ew, gross! Who'd wear that?"

She listed women at my Korean Saturday school who participated in this mutiny against good taste. Edward Kim mommy, Michelle Choi mommy, Sujin Lee mommy, to name a few. "You know, Anne, it cost lot of money."

I looked for the tag. "How do you know? There's no price."

"Because it *Veh-sahch-ee*. Not have price because people who buy not care price. They see ugly shirt—oh Mommy can't believe how ugly—and they see name and they buy even if three hundred dollar and they can—"

"THREE HUNDRED DOLLARS?"

It was the first time I had heard of such a thing, spending a large sum on such a revolting shirt, something made out of curtains pillaged from a yacht. Until that moment, I had assumed that hideous clothes cost nothing, because that is what I wore and that is how much they cost.

Even though my mother had refined taste, she did not use it when she dressed me. Instead, she dressed me the way my aunts dressed their children. Exactly like them. As in, I wore all their old clothes, which would've been fine had they just been normal. Because I was the youngest and smallest of all my cousins, my closet was the last stop on a hand-me-down's journey through time and two continents.

The odyssey would begin in Seoul, where my aunt would buy a new sweater for her twins Jung-ah and Jung-yun. In the beginning, they probably fought over the sweater, wanting to be the first to wear it. Then after a year, when the novelty had worn off, they most likely mashed it into the back of a drawer along with their mis-matched socks. Eventually the twins would unload the crumpled sweater on Yoon-chong. As the artist in the family, Yoon-chong probably modified the sweater, adding ribbons or pins, and she would wear it often, only to outgrow it and pass it to her younger sister. Yoonmi, as the free-spirited dancer in the family, probably ripped off the ribbons to put them in her hair and used the sweater as a jump rope until the stitches barely held together. Now the sweater, by this point faded and stretched to fit over the chang-ing bodies of four girls, would migrate across the Pacific Ocean to Tina in Los Angeles. The shipping alone would've cost more than a dozen new sweaters, but my family has never understood cost efficiency. At the command of her mother, Tina would wear the sweater until her arms outgrew the sleeves and the cuffs no longer covered her watch. By the time I got the sweater, it would stink of dried fish and mothballs, and have several stains—pickled cabbage, soy sauce, mustard, chocolate. There would be a series of holes in each armpit and the hem would unravel at the slightest movement. But these would be the least of its problems. Even after my mother would dry clean and mend the sweater, it wouldn't be suitable for wearing. It wouldn't even be suitable to line the bottom of my rabbit's cage. No amount of chemical solvents and thread could repair a sweater that was hopelessly out of style. Still, my mother would force me to wear it anyway.

According to my mother's rationale, it would be wasteful and dis-respectful to not wear these clothes after they traveled so far to get to me. Hand-me-down clothing was a way to keep in touch with our family in Korea, to hold on to the people we rarely saw. I'm sure my mother realized how atrocious these hand-me-downs were, but she figured that if these clothes were good enough for my Korean cous-ins, they were good enough for American me. She figured wrong.

At one point in Korea, any kind of English writing was cool when it appeared on clothes. When I was six, I received a gold, nylon vest printed with the words "The fun of soup bring Spring." My mother didn't know what it meant, and neither did I, but she did know that my cousins were hip, so if I wore their clothes I'd be hip too, a kind of secondhand coolness. This vest also had matching nylon pants with the same phrase printed along the entire side of one leg. The *shoop-shoop* sound the pants made every time I moved was a constant warning to all that could hear and read that I was an undeniable loser. As I was learning how to read, I discovered what nonsense I was wearing, and when I could read, my classmates could read.

"What does that mean?"

"I dunno."

"You think soup is fun?"

"No."

"Yes, you do."

"No, I don't."

"Soup's not fun, it's boring. You like soup, so you're boring. Ha, ha, ha, Annie Fannie Choi-boy likes soup! Chinese, Japanese, Indian chief!"

I went home and cried to my mother.

"Why you cry? You clothes match, see? Top and bottom go together. Everything match! It very nice set, Anne. It fashion in Korea!"

"But I'm not in Korea!"

"Anne, I tell you, if you in Korea, everyone say you look like model!"

"But I'm not in Korea! Everyone makes fun of me."

"Then tell them you model in Korea and they bad kid. Very bad. And they make they mommy such shame."

Another problem was that all the clothes I got were about a decade behind the current trend in Korea, which was yet another decade behind America in fashion. So according to my calculations, the hand-me-downs I was forced to wear in 1986, were actually from 1976, but looked like they were from 1966. So in the 80s when most of the kids were wearing neon-pink, peg-legged jeans, I was wearing dirt-brown bell-bottoms. When girls were wearing long, belted tops that fell off of one shoulder, I was wearing checkered polyester pant suits.

Then there were clothes that looked over two hundred years outdated. One of my mother's favorite outfits she forced me to wear was a pair of bright green knickerbockers and a white blouse with an unruly amount of lace ruffles at the collar and cuffs. It was an ensemble appropriate for Paul Revere's stable boy, and the first time my mother laid the outfit on my bed before school, I felt she had stopped loving me.

"NO, NO, NO! It's ugly, I hate it. I don't want to wear it."

"No, Anne, it cute. Mommy like it very much."

"Then *you* wear it!"

"ANNE!"

"I said, I don't want to wear it!"

"You have to wear, I make you wear."

"I want to wear something else. THIS IS UGLY!"

"Anne, shhh, everyone like it. You teacher say 'Oh Anne you so pretty today I wish I had same clothes!'"

She shoved me into the car, with my frilly blouse and knickerbockers soaked with tears. She assured me that I would be the center of attention, and she was right. My eight-year-old classmates stared until their eyeballs popped out of their heads and even my teacher looked confused. Miss Jensen didn't know how to tell me that the American Revolution was over.

Another one of my most memorable garments was a heavy wool, dark brown dress with a white lace collar and an oversized velvet bow. The dress came nearly to my ankles and looked as if it belonged in *The Little House on the Prairie*—all it needed was a bonnet and Michael Landon. My mother found this outfit so adorable that she forced me to wear it in the summer. On a 110-degree Southern California afternoon, the wool dress was not appropriate. I itched. I sweat. I cried. Then I took it off when she wasn't looking. Once this was at the grocery store. The ice cream SECTION OF OUR GROCER'S FREEZER WAS JUST BEGGING ME TO STRIP.

"ANNE! What happen to you dress?"

"I don't know."

"WHERE IS DRESS?"

"It's too hot, it's too itchy. I hate it."

"YOU PUT DRESS ON NOW. YOU RUN IN STORE NAKED. I SO EMBARRASS. YOU MAKE MOMMY VERY MAD!"

My mother, so dramatic—It's not like I was naked. I started wearing an undershirt and shorts underneath the dress in order to minimize the contact between my skin and that itchy, burlap sack. This probably contributed greatly to my overheating, but choosing between a heat stroke and a rash was not easy.

When I turned nine years old, I started picking out my own clothes for school, and I salvaged a few articles that were possible to wear without being stoned by classmates. After years of negative reinforcement, I had figured out which clothing was acceptable. Miraculously, Tina had passed on a pair of plain, white pants, and I was finally making friends. These white pants were unobjectionable, unlike the red and white polka-dotted square-dancing dress with the fifty-pound pink petticoat ("But you love pink color!").

My mother and I started shopping to buy clothes to supplement the ones from my cousins. Everything I picked out in the store was a solid color—white, beige, gray, black. The wide-legged, paisley jumpsuit didn't look so bad after a blue jacket covered half of it up. At last, pants and shirts and dresses, absent of extra zippers and large buttons that served no purpose, made me feel at peace.

When I started middle school, the trickle down of clothes from my older cousins finally stopped. They had all gotten their growth spurts, and their broad shoulders cruised at a higher altitude than mine. By twelve years old, I still measured less than five feet tall and weighed well under a buck. But even though I looked more like an eight-year-old than an eighth grader, my mother decided that it was time to dress like a sophisticated lady like herself.

"What you think this skirt, Anne you like? It tweed."

"Ugh, Mom, It's like totally itchy. OH MY GOSH, it has a matching jacket TOO?"

That woman pushed tweed suits like a drug. At the junior's department at Robinson's, my mother could sniff out tweed in a rack full of cotton separates. She didn't understand that most junior-high girls just wanted to fit in, which meant looking like other junior-high girls, not Jackie O. or a basement couch.

"OK, ok you not like it. How about this, you like?" She held up another skirt.

"Mom, that's tweed too. *Stop it.* You're totally embarrassing me!"

By twelve years old, I had developed my own sense of style, which was to lack it. I wore nothing that would call attention to myself. I refused to follow any trends—I was afraid I would follow them incorrectly or at the wrong time. I had spent enough time under the scrutiny of my peers, and I just wanted to remain under the radar. No more confusing translations displayed proudly on sweatshirts, no more anachronistic clothing, no more jumpsuits.

"Anne you look so boring. You wear same thing every day, shirt and pants and tennis shoe. I fall sleepy when I look at you."

"What are you talking about? How about this?" I picked a shirt off the sales rack.

"Anne that has no style."

"Sure it does, it's black and has a collar. Pretty stylish right?"

"Oh, my only daughter look like boy! I think maybe I die!"

Shopping had become an exhausting, exasperating routine for us. She wanted me to dress with finesse, to be a fashion-minded daughter of a fashion-minded mother. It was important that I looked good now that I was older, partly because it reflected on her parenting skills. Daughters who are stylish are organized, obedient, and Ivy League-bound. Daughters who look like boys are indolent, rude, and start fires at school.

"Anne, why you not wear dress?" She held up a yellow dress with a lace skirt.

"Because I don't want to wear an ugly dress."

"What you mean? Dress is pretty!"

"No it isn't. It looks like a tablecloth. There are so many dresses here, and they are all like totally ugly."

She held up a flower-print dress, which I deemed childish. She held up a black dress, which I called depressing. She held up a white lace dress.

"You're kidding, right? Am I getting married?"

"Anne, I think no one marry you. You have so much excuse! You make Mommy life very hard." She held up a simple, blue T-shirt.

"Oh nice, I like the pocket."

She rolled her eyes and took out her wallet. Even if she bought me a tweed suit or a wedding dress, there was no way I would wear it to school, or anywhere else for that matter. I was too old for her to dress me, and she was too tired to argue with me about clothes.

In high school, a petite button-nosed girl named Alyson Spilker introduced me to vintage stores. Alyson had blue hair, a nose ring, and a quirky sense of style that I admired. She wore outlandish pants, colorful hats, and big silver boots. She made her own shirts out of tights and created her own jewelry from wires and beads. Alyson was charismatic and charming, and as we became closer, Aardvark's Odd Ark and Hidden Treasures replaced my mother's cavernous, cloned department stores. I discovered that shopping could actually be pleasant, and I realized that used clothing from other people—as long as they weren't my cousins—could actually look good. Alyson and I would paw through dusty clothes, and she would always find the most misshapen dress or the most chaotic sweater in the store and laugh.

"Oh my God, can you even imagine wearing this?" She held up a purple sweater-dress. "It's like someone was knitting a sweater and said, hey, I wonder what'll happen if I just keep going?"

"I've worn worse things."

Through Alyson, I developed a style of my own—sequined sweaters from the fifties, geometric scarves from the sixties, coats from the seventies, and select pieces from the eighties. Hiding in plain shirts, pants, and tennis shoes wasn't necessary as I gained more confidence. I never wore anything too eccentric, only clothes with just enough inventiveness to make me feel comfortable and noticed without feeling out of place. The clothes were unique and affordable—sophistication at a sensible price. I guess I did learn a little from my mother.

"Anne, why you always wear old clothes? Why not buy new?"

"Because the old stuff is cool."

"But it old, you look like homeless!"

"No, I don't. Homeless people wear trash bags."

I foraged in my mother's deep closets for her old clothes, finding blazers with patches on the elbows, macramé belts, and printed blouses with long sleeves that I had my grandmother shorten so they'd fit better. I found daring mini-skirts, fuzzy cardigans, and even a leather trench coat with a faux fur collar.

"I don't like you clothes, Anne. We go shopping more. You look so silly."

"But these are *your* clothes. So are you saying that *you* look silly?"

"Anne, I tell you, I wear those many, many year ago. Before you born and give me headache."

"Then why did you keep them around?"

"I don't know. I should have throw away but I think so much waste."

"OK, so now I'm not wasting them, what's the big deal."

"Big deal? You look silly—that big deal. People think 'Oh Annie mommy not dress her only daughter so Annie have to find old clothes.'"

"They so totally do not think that. They think, 'Oh Annie's so cool. Look at her awesome clothes. I wish I was wearing that.'"

My mother couldn't help smiling. It is a strange compliment, that someone could appreciate the sense of style you had decades past instead of the one you had at the moment. But fashion changes, and as trite as it may sound, people change, too. And people's fashions change. And sometimes this leads to crimes of fashion.

When I started college, my mother started golfing. With no kids in the house to obsess over, she quickly settled into a routine. Every morning, she practiced at the range with friends from church, and got lessons from a pro every Thursday. At least three times a week she played eighteen holes, sometimes thirty-six if she

#216 05-06-2012 1:51PM
Item(s) checked out to p32064226.

TITLE: The night trilogy
BRCD: 30641003504771
DUE DATE: 05-29-12

TITLE: Happy birthday or whatever : trac
BRCD: 30641002736788
DUE DATE: 05-29-12

could squeeze it into her day. Within three months, she bought a set of titanium clubs and high-end golf balls designed to fly to the moon. Then she enrolled in a country club. She would call me on the phone each week to talk about the newest trends in putters or the latest improvements to golf cleats, and I'd sigh loudly hoping she'd realize she was boring me. Like fashion, golfing was something that put her in a special group, but to this I just couldn't relate. I thought golfing was for people who were so wealthy that they had nothing better to do than chase a little white ball over some hills. Golfing was for the rich and white. My mother was neither. Her golf clubs were on layaway and her first country club was right next to the 101-freeway—large nets prevented balls from smashing through windshields. To my mother, golfing was the next step up. Her stylish clothes made her look sophisticated and wealthy, but being good at golf was a way to act sophisticated and wealthy. Plus, many of her church friends golfed. She needed to keep up.

The first Thanksgiving after starting college, I returned home and noticed a few awards on the shelves. In just a few months, my mother had become quite a decorated golfer—there were a certificate of participation for a church golf tournament, a Booby prize for the worst but most spirited player, and an award for Longest Drive. Her awards were next to me and my brother's high school diplomas, a trophy Mike got in Boy Scouts, and a plaque awarded to my father by his company for his dedication and commitment to excellence.

"What did you do with my fourth-grade photo?"

"Oh I put away to make room for Mommy trophy." She waved her hands casually.

"What? Why would you do that—whoa, what happened to your hand?"

One of my mother's hands was golden brown. The other was pasty white.

"I wear my golf glove on this hand."

"It looks like you're still wearing the glove. It's freaky."

"What mean freaky?"

"You know, like scary."

"You scare? I scare too. I scare that my only daughter look like homeless." She looked over my outfit and recoiled. I was wearing a vintage leather blazer, a light blue tuxedo shirt, and jeans.

"Whatever. Why don't you wear sunblock?"

My mother grinned and proudly showed me her shirt tan. A perfect line circled each bicep, marking a border between her bronze arms and her pale shoulders. It was the most shocking farmer's tan I had ever seen, but there was something much more shocking—her outfit. She was wearing blue and green plaid Bermuda shorts with a red and yellow plaid collared shirt. On her tousled head was a terrycloth visor with plaid trim, also clashing. She was a confusing map of horizontal and vertical lines, in more colors that should be worn on one person. Her thick white socks had gigantic pom-poms on the back, which prevented the socks from slipping into her golf cleats. Her feet looked like she had stepped inside two cottontail bunnies. Her outfit was worse than *The Little House on the Prairie* throwback dress and more ridiculous than the Paul Revere ensemble. But not worse than the "Fun of Soup Bring Spring."

"Did you golf today?"

"No, not today."

"You didn't? Then why are you wearing the visor?"

"Because it part of set."

"That's a set? But you said you didn't golf today."

"So?"

"So, why are you wearing golf clothes? Why aren't you wearing normal clothes?"

"This normal clothes, Anne."

My mother, the one who used to scrutinize Korea's version of *Vogue*, the one who taught me the difference between Kenneth Cole and Cole Haan, now looked like she shopped exclusively at Golfsmith and singlehandedly exhausted all the plaid in Scotland.

"Mom, are you going to change before we go out to lunch?"

"I did change."

"Well don't you think you should put on a dress? Something pretty? You can't go out like that."

My mother exploded with laughter. She grabbed my arm and clutched it to her chest, shaking me. The plaid on her clothes quivered, giving me mild vertigo.

"What? What's so funny?"

And then it occurred to me that I sounded like my mother.

My face turned into a gigantic eggplant, for what child wants to sound like a parent? I was even holding back. I had nearly asked my mother to clear her crap off the kitchen table, which was cluttered with old mail, church newsletters, phone books, and a pile of muddy golf tees.

I rolled my eyes and left my mother chortling in the kitchen. I walked into the master bedroom and took a peek into my mother's closet. I was aghast. I saw what seemed like hundreds of collared shirts, in plaids and stripes and even animal prints—tiger and zebra. I wondered if my mother ever mixed prints, so predator and prey could meet on the lush green hills of Los Robles Golf Course. Her drawers exploded with overly pleated shorts in a dizzying array of oranges, greens, purples, and puce. Her stylish dresses and skirts were pushed aside and crumpled to make way for windbreakers, golf slickers, and sweater vests. I shook my head. My mother has always been fashion-savvy, so what if hers were the best threads on the links? What exactly were the badly dressed golfers wearing? I shuddered.

I picked up a shopping bag from a pro shop off the floor. It contained an extremely chunky green wool sweater with a giant appliqué of golf clubs and a ball. The price tag was still attached. I sprinted back to the kitchen.

"MOM! HOW COULD THIS COST ONE HUNDED AND TEN DOLLARS?"

"Anne, it on sale!"

"Are you out of your mind? No one should pay for a sweater like that. It should be free."

"It style! It Callaway, very famous. Make golf club and clothes."

"Well there's the problem. Companies that make golf clubs have no business making clothes."

When did my mother forego sensible, silk blouses for wallet-gouging, wooly mammoth sweaters? What happened to that elegant lady? And, more importantly, why was she wearing a visor indoors? Golfing had ruined her fashion sensibilities and my eyesight.

"Anne, you ready to go lunch?"

I flinched at her outfit. I had a choice here: I could force my mother to wear something civilized, or let her be her and let me be me, which in this case would be an accomplice to the worst fashion felony since 1982, when my mother forced me to wear a puffy barber-pole-striped dress that had matching pants attached underneath. This outfit, meant to offer the femininity of a dress with the safety and comfort of pants, put me in tears because I couldn't figure out how to take it off to go to the bathroom. The surly hag of a recess aid—the one everyone feared—had to help when she saw me jogging in place with my hands cupped around my crotch.

"Yeah I'm ready, but can you take off the visor?"

"No, I tell you, it part of set."

"Take it off."

"No, Anne, I say no."

"You can't wear it. I won't let you. I can't eat with you if you're wearing a visor."

"Anne, why you make Mommy angry?"

"Why are you wearing all that plaid?"

"Why you get you clothes from trash can?"

"This jacket is yours."

"No I think you wrong. I never see this jacket."

"OK fine I bought it at a store, but come on, take off the visor. Do that."

"Anne, NO."

We ate at California Pizza Kitchen and my mother babbled loudly about golf—she had just volunteered to organize the next church tournament. I hunkered in the corner of our booth, hoping the power would go out.

stroke order

Although my brother and I were born in America, Korean was our first language. My parents never taught us English because they only had a small working knowledge of the language—hello, excuse me, thank you, how do I get to the 101-freeway? By the time I was born, my parents had been living in the States for five years, and they had found ways to work around their limited English. As a chemist, my father spoke and wrote in the international language of elements, compounds, and formulas. My mother had taken English classes at a community college, but she didn't practice her skills much because she sought out other Korean immigrants. She didn't need English to buy groceries or drop off dry cleaning or get a haircut—merchants in Los Angeles's growing Korean community offered goods and services at prices lower than their English-speaking counterparts with none of the embarrassing hassle of staring blankly at labels and faces. Though their English has improved considerably since immigrating thirty-five years ago, my parents still struggle with the language today. Whenever I watch a movie with my mother, she tugs on my sleeve every ten minutes and asks me to translate, not into Korean, since I'm incapable of that, but into a simpler version of English.

"What happen?"

"Leonardo DiCaprio is in love with the girl, and other guy doesn't like it."

"Which one Leno Decrap?"

"The short, blond one—yellow hair—with the big head."

"Why that man tie Big Head to table? Why they fight?"

"Because he's in love with the girl, too."

"I not understand—why they fight *now*? Boat sink, who care? Everyone drown and they fight? This movie so silly."

"Mom, shh, there's still another hour."

"Hour? Oh my gosh, how? Boat sink in ten, fifteen minute!"

Befuddled by the rules of English and all of its illogical exceptions, my parents figured that English should be something that my brother and I learned in elementary school from trained professionals. The rest of the family shared this sentiment as well. My cousins, who knew only a mouthful of English words when they immigrated, were dropped into the American school system, and they figured it all out eventually,

My mother tells me that as I child I was shy at first, but after I warmed up I was quite chatty. I talked to my relatives in Korean, played Korean games, and sang Korean songs. Even my stuffed animals spoke Korean to each other. My first word was *ohm-ma,* or "mommy." Korean was what I knew. But then I entered elementary school and suddenly what I knew was not what everyone else knew.

On the first day of kindergarten, my mother took me to my classroom and handed me my lunch. I sat down, ate it, and in five minutes I was ready to go home. When Mrs. Smith began talking, I realized that not only was my mother nowhere to be found, but also I had no idea what this stranger was saying. Just like a few other children, I started wailing, "Where's my mommy?" but I cried it in Korean. Of course, no one understood, but Mrs. Smith figured out quickly that I didn't want to be there. I wept, sputtered, and sniffed and I kept on looking at the door, expecting my mother to rescue me like she always did whenever I had a major freak-out. When Mrs. Smith bent over and put her arm around me, I screamed. Mrs. Smith was shaped like a pear and had tight dark curls that set off her pale, ghostly face. She wore a large silver stopwatch around her neck and had long purple fingernails that curled menacingly. This was not someone I wanted comfort from; I wanted someone who smelled like garlic and sesame oil and had delicate, thin fingers with trimmed nails.

Mrs. Smith took me to a room with long tables where young children squirmed on one side and adults sat patiently on the other. There were pictures, blocks, and spiral notebooks littered between them. The school administrators gave an IQ test to every incoming kindergartener—in English. Even though it was clear that I didn't understand the language, I still had to take the test to prove that I didn't understand the language. I don't remember many specifics of the test, but I do remember one particular question the tester asked me.

"Which picture is a steeple?"

He held up three pictures and motioned for me to point, repeating the word *steeple* rather loudly, the way some people do when they talk to a non-English speaker, as if increasing volume will somehow increase comprehension. My choices were an American flag, a group of stars, the top of a church, and a tree. I chose the flag. Even if they had asked the question in Korean, I'm pretty sure I would've gotten it wrong. I wasn't going to church yet and I had never seen a steeple before, at least not the kind white people put on their IQ tests.

If one could fail an IQ test, I suppose I did. School administrators labeled me "special," and every morning my mother sent me to Mrs. Smith, who then ushered me into a "special" class just for kids who didn't know what a steeple was. I was in a remedial learning class because the school didn't have an English as a Second Language specialist. My brother and I were the only two non-English speakers in the school. We were also the only two "Orientals." (Actually, I'm wrong, there was also Otto Ho, but he was fluent in American things like Mexican food and *Three's Company*, so he didn't count.) I joined the remedial learning class half of each day, along with a mentally challenged girl with enormous glasses that made her eyes look like gigantic russet potatoes. We didn't really belong together, since she spoke English and I didn't, and I could button my own clothes and she couldn't, but this is where I learned how to speak, read, and write.

Despite the language barrier, I still managed to make friends. The nice thing about kindergarten is that you can play with other kids and don't really have to talk. You can just jump rope, swing, and get sand in your pockets. Occasionally you throw a fit or whine when you want something. But no kindergartner ever asks if you have a gong or came over in a boat or eat dog—all of that comes later. But as I made friends, my mouth began shaping English words into mostly coherent phrases. And that was the beginning of the end, as it were.

Children learn languages quickly, and English began replacing my Korean. By first grade, I spoke in Korean and any word I didn't know, I just replaced it with the English word. Whenever my mother asked me, in Korean, what I did in school, I would answer something like this: Korean, Korean, merry-go-round, Korean, slide, Korean, Korean, fingerpaint, Korean, Korean, sandcastle. My parents would teach me the Korean word, but I ignored it. There was no sense in learning the same word in two languages, right? I figured as long as my family knew what I was talking about, it didn't really matter which words I used from what language.

My brother and I attended a mostly white school in a mostly white suburb far removed from Los Angeles's Korean community, which was downtown and afflicted with downtown problems like crime, homelessness, and substance abuse. Our neighborhood had parks, crossing guards, and mountain trails afflicted with mountain problems like cacti, ticks, and poison oak. Everything around me was in English—classes, books, television programs, menus, signs, labels, voices. By second grade, I had absorbed all that English and spoke mostly in English to my family. I did, however, sprinkle sentences with a few Korean words. When my mother asked me what I did in school, I would answer something like this: I sang a *noleh,* played handball with my *chingoo,* and drew a *geleem* of our house. I had a lot of *jehmee.*

My parents still spoke to me in Korean, and I understood them perfectly, but I answered in English because it was easier. English, not Korean, was the language in which I now thought and dreamed. At seven years old, I was exploring language like my classmates. I enjoyed English tongue twisters, puns, jokes, and tall tales. I liked listening to poems where English words created rhythms and patterns I hadn't heard in the language of our home. The Korean versions didn't interest me because I had no one to share them with except my family, and what was the fun in that?

My parents recognized my waning Korean as a problem. First, my mother showed me videos of Korea's answer to *Sesame Street*, which I thought was boring because what I really wanted was *Sesame Street*. Then, she gave me Korean picture books, which also failed to keep my attention. My brother, whose Korean had also degenerated, rejected all of his Korean books and preferred his *Hardy Boys* and *Star Wars* serials. Eventually my parents realized it was time for more formal measures to make us read, write, and speak Korean like Korean people. They enrolled Mike and me into a Korean school held on Saturdays, and really, that is what all children really want—more than a trip to Disneyland or a Golden Retriever puppy—a sixth day of school.

The Korean classes were held at a junior high school about forty minutes away from where we lived. I felt odd sitting in such a foreign classroom, one without colorful banners, watercolor artwork, and an aquarium with caterpillars. Instead, there were D.A.R.E posters, large maps, and diagrams of the human body (on which many students, including me, pointed to the crotch and giggled). Though I was in second grade, my parents enrolled me into the first grade of Korean school. I had to start from the beginning, they explained to me. My brother, being three years older, was in a class designed to teach Korean to older students—the pace was much faster and the homework load much heavier. Most of my classmates were my age, in the same situation—they too had become more inclined toward English than Korean. My teacher was a middle-aged woman who pointed to students with a ruler and occasionally slapped it on a desk when we answered incorrectly. She was short, had greasy skin, and her hair was a gigantic, frizzy mess. She looked like a Korean troll and for a while I was convinced she lived under a bridge, or perhaps one of the freeway overpasses. She passed out large sheets of paper and had us copy the Korean alphabet over and over and over again. And then over again.

The Korean alphabet doesn't have an l or r sound as Americans know it. The closest sound is somewhere between l and r. Oddly enough, it is similar to the t/d sound in the word *water* (when it is not pronounced in the enunciated Martha Stewart way—"wat-ter"—or in the garbled Philadelphian way—"werddur"). The sound of this Korean letter is subtle. As air moves out from the throat, the tongue gently flicks the roof of the mouth, behind the front teeth. It's like a quiet purr or a gently rolling r. However, the letter, called *lee-ul* (maybe *ree-ul?*), looks nothing like the sound. It looks like a backward s, but instead of a curving, serpentine shape, the Korean letter is written with horizontal and vertical lines. It's like the 2 on a digital alarm clock, the kind with the red numbers that burn into your retinas when you can't fall asleep.

When I first learned how to read Korean, this letter deceived me. I pronounced it like an s, as in *strenuous*, not like the t as in *daughter*. It was confusing and part of the reason why I got held back in Korean school. At the end of my first year, my Korean teacher told my mother I was reading at a kindergarten level. I didn't even know kindergarteners could even read—apparently they could in Korean school. So, I was forced to take the Korean first grade twice.

"Anne, you make Mommy so mad, *weh gongboo ahn heh?*"

"I do too study! I *gongboo* everyday! It's my *sonsengneem*. She's so mean. She doesn't like me."

"Always excuse. You fail and now you take *eelhanyong* again. *Ayoo*, Mommy *hanadda!* So shame!"

"I hate *hangook hagyuh!* I don't want to go—*ahn ga shipoh!*"

"No, you say wrong. Say *ahn gagoo shipoh,* not *ahn ga shipoh.*"

"I don't care. I'm not going!"

"Anne. You. Have. To. Go."

Somehow my mother and I understood what the other was saying, though we didn't really listen to each other. Because my mother's English and my Korean had large gaps in vocabulary, we developed an awkward Koreanglish. We pulled words and phrases from both languages and transitioned so seamlessly that I often had no idea what language we were speaking, or yelling. Sometimes, in arguments where my mother yelled at me completely in Korean, I responded completely in English. It was a power struggle, a battle for turf. She felt more comfortable in Korean, and wanted me to argue in Korean, where I would have a distinct disadvantage. I, of course, wanted it all in English for the same reason.

The second time around in the Korean first grade, I got a teacher who focused on "having fun." She drew silly pictures to teach words and gave us word searches and comic strips, but no matter how enthusiastic she got about vowels and consonants, I hated learning Korean. It was infuriating enough having to go to "regular" school Monday through Fridays, learning the increasingly complex rules by which numbers and English words operate, but on Saturdays, I had to learn a whole new set of rules for a whole new alphabet, while my "regular" friends went to birthday parties and sleepovers and played Atari. Then on Sundays, I had to go to Bible school. I went to school seven days a week. I also attended dance classes and piano lessons, practiced piano everyday, and had to choose between *swam* and *swimmed* on my Mommy homework.

The intensity of my schedule caused my grades in "regular" school to slip, and I received B's and B+'s on my tests, which horrified my mother. Though learning Korean was important, my

American education was the priority. Despite my protests, my mother cut my dance classes to give me more time to study—I had preferred to cut piano instead. Still, by the time I got around to my Korean school homework, I was just too exhausted to read and write sentences and memorize vocabulary. And when it came time to read a book where Jesus appeared as a cartoon figure and write a paragraph about why I was blessed, I fell asleep. Fortunately, Bible classes didn't have grades, but my teacher told my mother I couldn't even name all the commandments, which was true but, really, my mother didn't need to know that.

"Why you such bad student? Everyday I worry that you fail. Then you not get job and you get hungry. Then you get cold and sick. Anne, you not study, you die. Why you give me heart attack?"

"I go to *hagyuh* everyday! There's too much school. I can't keep up. It's the stress." I was very proud that I knew the word *stress*. It made me sound like an adult.

"What? No, no, Anne, you not get stress. Mommy get stress! You fail spelling and God and *hangook* school, Mommy get so stress!"

"Why do I need to go to Korean school? I speak Korean!"

As an eight-year-old, I saw no need to go to Korean school and learn letters and words I didn't care about and had no need for. As a mother who saw her daughter lose the tongue of the mother country, she felt concerned. If I couldn't read, write, or speak Korean, then what kind of Korean would I be?

I finally managed to pass the Korean first grade, but shortly after I entered the Korean second grade, my teacher approached my mother—even after two years of Korean school, I still had difficulty reading and writing the alphabet. My mother brought home Korean workbooks, written for the grade level below me, and with the watchful eye of a prison guard, she supervised while I practiced writing Korean letters on large sheets of grid paper. As I painfully copied a letter in each box, my mother showed me the stroke order in which to write each letter, something I had always ignored.

In English you can write an E however you like, and no one will stop you; it's truly a free country. You can write the vertical line first, and then draw the three horizontal lines, or vice versa. You can even draw the top line, the vertical line, the bottom line, and then add the middle line. In Korean, there is an emphasis on stroke order. One must draw from top to bottom, then left to right, even if it's not efficient. For example, this is how you are supposed to write *lee-ul*, the letter that looks like a backward s.

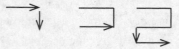

To me, this was incredibly inefficient. My mother explained that this stroke order emerged because Koreans used to write with calligraphy brushes, and now it was just tradition.

"But I don't write with a brush. I write with a pencil. So can't I just write the letter however I want?"

"No, Anne. There is order. This only way!"

"But Miss Jensen said there was more than one way to do something!"

"Miss Jensen not Korean. In Korea there only one way!"

Only one way? Well, I thought, that is just wrong. Obviously you can write the letter in one swift motion, without having to pick up the pencil off the paper, and it will look exactly the same. So, whenever my mother wasn't looking, I wrote all the letters the way I wanted to write them—whatever was easiest and fastest.

As my parents got busier—my father with work and my mother with church—I began doing my Korean homework unsupervised. Immediately my handwriting got sloppier, and my lee-ul started losing its 90-degree angles. It mutated into a curvy backward s, which only added to my confusion when I read. As my Korean teacher pointed out, my handwriting was so illegible it was as if I had a hook for a hand. My ignorance of stroke order, she assessed, was my problem. But this is the way I still write *lee-ul* today, in those rare cases I write in Korean. I had to look up the stroke order on the Internet.

I attended Korean school for six years and eventually I learned how to read, but my penmanship never improved. The Korean alphabet is full of straight lines and geometric shapes and right angles, and my letters looked soft and squishy and bloated, just like a fat, lazy American. My vocabulary never really improved either and to this day I speak like a third grader, using short words in simple sentence constructions. I get quiet and shy when I speak Korean around my parents' friends and colleagues; I fear that they will judge me based on my Korean skills, which will reflect poorly on my parents. Even though my parents can drive me into a blinding postal rage, I still feel sorry for letting them down.

As I've gotten older, I've made more of an effort to speak Korean to my family. Whenever I see my cousins, I try to carry conversations in Korean and they always speak to me in English because they want to practice their skills and learn all the slang I usually throw around. Lately, I've been watching more Korean films; one of my favorite directors is Park Chan-wook, whose films are much more violent and disturbing and heartwrenching than any American movie I've ever seen. When I visit my mother, sometimes I'll watch Korean soap operas with her and she'll translate for me.

"What's going on?"

"Both sister love same man."

"Who does he love?"

"Old sister, one with long hair. But her husband die so she not ready, but everybody know she love him. She tell him to marry sister."

"That's stupid. He doesn't love the sister, why would he marry her?"

"Because he love old sister and he do what she say."

"*Booorrring*. Korean soap operas are just as lame as American ones."

"Anne, quiet, I watch!"

"But this is horrible. Change the channel."

"No, I like."

"Let's watch that game show."

In one of my favorite Korean game shows, contestants must execute rather embarrassing and physically demanding tasks. In one episode, several groups of students had to wear 80s athletic gear—sweatbands, striped knee-high socks, tiny polyester shorts—and hold on to ropes hanging from the ceiling. Then, they had to answer random questions for points. (What year was our president born? Which weighs more, a monkey or man in a monkey suit?) The team that hung on to the rope the longest and had the most points won a prize that wasn't worth the effort, maybe a coupon or something.

"Anne that show stupy. Finish this one first. Then we watch different."

"Fine, fine, but this is really bad. Oh look, now everyone's crying."

Three years ago for my mother's birthday, I sent her a card with a black-and-white photograph of a large group of nuns. Their backs were facing the camera, with their dark habits draped generously over their heads and shoulders. One pale, wrinkled nun was turned around, sternly shushing a gleaming white crying toddler, who was wearing nothing but white diapers. Normally I would have written birthday wishes in English, but instead I wrote a short message in Korean. I wanted to show my mother that now, after many, many years, her daughter was trying to reclaim the language she once knew and then forgot and then rejected. I suppose it's pretty ironic: I began life being "special" in the English language and now I'm "special" in the Korean language.

I overemphasized the angles and lines of each Korean letter, making sure each right angle was right and each line was perfectly straight, though I admit to ignoring the stroke order. I wrote, "Mother, you are very old! I am your only daughter. Annie."

My mother called me in a fit of laughter, saying that my handwriting looked like a first-grader's, the way a kid pays so much attention to how the letter looks that it is devoid of any style or personality. My mother felt that this was especially amusing since style and personality were not things I lacked. Then she explained that I misspelled the word *mother*.

"Oh my gosh, you spell like baby!" She laughed loudly over the phone.

"You know, I can play this game, too. What do you study when you learn about Socrates and Plato? What do you call that?"

My mother started laughing even harder.

"Come on, tell me, what did Socrates study?"

"Anne . . ."

"Come on, say it. Say it for me."

"Pflipspy?"

"What? No you're close . . ."

"Pflospsi? Oh too hard!"

I laughed so hard that I snorted. My mother's mouth has always struggled with the f and l sound, something that doesn't occur in Korean.

"OK, OK, good enough. What color is a plum? Or grapes?"

"Oh so easy, Mommy can do this one—puppel! My only daughter think she such genius—how you say 'my room'?"

I laughed and then stopped—wait, how do you say "my room" in Korean? I always got "my room" and "moth" confused. One was *nahbang* and the other was *nehbang*. To my undistinguishing American ear, there was barely a difference. For years I always said, "I'm going to my moth!" or "I know my moth is messy!" or "Stay out of my moth!" My mother thought this was sort of cute—it was a mistake many young children made, even though I was saying it when I was seventeen. Even my relatives thought it was adorable, although a little pathetic. Everyone always corrected me. Annie, they explained, you do not sleep in a moth. But I always ignored them.

"No, no I know this one, I know this. It's . . . *nahbang?*"

"Wrong! You got wrong!"

"No, no I meant *nehbang.*"

"Sure, sure. Anne, you should have study in *hangook hagyuh.*"

"I know, I know, you're right. I should have study."

period piece

To say that I was a "late bloomer" is like saying childbirth is mildly uncomfortable. Doctors claim there's no "right age" for a girl to start menstruating, but every girl knows the truth—getting the first period too early or too late can do more social damage than farting explosively during study hall. Even the term "late bloomer" itself insinuates that there's a scheduled time, a planned date and hour, like a manicure. Most girls get their first period when they're eleven or twelve years old. I was seventeen when I got mine.

I applied to college before I was a woman. I had my first job before I was a woman. I even learned to drive before I was a woman. Imagine, if you will, an enormous cruise ship helmed by a ten-year-old. That was me, driving my mother's white Cadillac at sixteen. I was four feet, ten inches and weighed ninety pounds soaking wet, and my main challenge was not parallel parking, but getting the seat high enough to look over the dashboard, but low enough to reach the pedals. Moving the chair somewhere in the middle meant watching the road through the spaces in the steering wheel. My driving instructor took one look and knew he was walking into a deathtrap.

"You're sixteen?"

"Yes. Yes I am."

"Have you considered learning on a smaller car?"

I learned on the driving school's compact Japanese sedan, which suited me better than my mother's American yacht.

Blooming late is in my blood. My mother got her first period at sixteen. Throughout most of high school, she was a pale, pocket-sized schoolgirl who feared any animal bigger than a bowl of rice.

"Mommy was so small. I wear green clothes I look like . . . like . . . lepra . . . leper. . . . Ah, you know small, green . . . very small . . . from Irish?"

"You mean a leprechaun?"

"Yes, Mommy look like leperka just like you!"

"I do not look like a leprechaun."

"When you baby you look like leperka, or little Santa Claus."

"What? No, you're totally confused. Santa Claus is big."

"No, no, like Santa Claus maid."

"Mrs. Claus?"

"No, you know, little maid, they make toy."

"You mean ELVES?"

"Yes! You look like elves!"

As a teenager, my mother cowered under Chang-hee, her taller, brawny sister whose voice sounded as if she chain-smoked Brillo pads, even when she was seven years old. Together they attended school in Seoul along with their younger sister, Jin-hee, who was mildly retarded from a birth complication, and their younger brother, Jae-sung, who walked with a limp caused by polio. Classmates teased the Hong children mercilessly and many walks to school were spent holding back tears and looking anxiously over their shoulders.

"They say Jin-hee and Jae-sung ugly and stupid. I so mad but what I can do? Mommy smaller than you! Chang-hee big, but she never fight. She such turkey."

"Chicken. You mean chicken."

"No she bigger than chicken. She turkey."

Apparently, Chang-hee was overwhelmed with the embarrassment that only retarded, crippled, and midget siblings can bring.

"Then I grow tall and everyone so small! I say, 'Why you yell at Jin-hee? She quiet and not bother you! You terrible!' Then I get taller and taller and they get scare."

After she hit puberty at sixteen, my mother soared to five feet, six inches, which was extremely tall for a Korean woman in the 1960s. It's actually pretty tall for a Korean woman now. In high school, my mother was quite athletic; she dominated her high school's volleyball team. I had assumed that when I hit ninth grade I'd do the same; volleyball looked fun, especially when one was on the winning team. But I quickly realized my mother's sport wouldn't be an option for me unless officials lowered the net by three feet and Reebok made platform sneakers. Basketball was out, too, even though my mother was convinced I'd "look so cute in basket uniform with big short pants."

I decided that athletics wouldn't be part of my high school scene, but my mother insisted that I play a sport—it'd look good for Harvard. I picked track and field because I figured that playing a team sport would've been unfair to my teammates, and with track I figured that if and when I lost a race, I'd only be hurting myself. Plus, the track coach never turned anyone away—anyone with two legs could run, he said. I had two legs and I could run. It just took me a long time.

Puberty is like any fad; it is always dangerous to be too early or too late. Take, for example, the 1980s trend of wearing gym shorts on top of sweatpants. I imagine that the first individual to make this bold fashion statement—that wearing underwear on the outside of your pants is cool—was brave and audacious, but probably suffered a ton of justifiable ridicule. But then the fad caught on and department stores started selling sweatpants with gym shorts sewn on top, and everyone enjoyed being trendy and comfortable and dumpy all at the same time. Anyone who caught on to the trend too late might as well have sported lederhosen.

Since my mother and I were late bloomers, we wore the lederhosen. My friend Christy Roland bloomed too early, so she was freakish and uncool, too. When we first started seventh grade, most of our classmates hadn't developed yet and they saw her as a grotesque anomaly, a husky girl with big plastic glasses and ample hips. Girls whispered behind her back and walked briskly past her, glancing curiously at the mounds in her sweaters. But then Christy became accepted, or perhaps tolerated, because everyone had finally caught up with her. If there's one lesson I learned from my high-school career counselor, it is this: Better to be early, than late. Miss Osborne was referring to job interviews, but I think the lesson applies to everything in life, with the exception of arriving at your own surprise birthday party.

Christy and I both played flute in the middle-school band, which my mother forced me to join even after I explained that band was socially "the worst idea ever." Christy had started growing breasts and "special hair" and all that good stuff at ten years old; I was a foot shorter and her nipples were at my eye level. Christy ignored most of the girls and instead eyeballed the boys, who didn't really know what girls were, other than mysterious and elusive animals that cried a lot and traveled in packs. She caught the attention of older boys she met on the street, in the Tower Records parking lot, and at the mall food court. By the time she was twelve, she was dating sixteen-year-olds who sold weed and bootleg Mötley Crüe tapes out of their trucks. During band class, Christy would pull me aside and recount her adventures in romance.

"My boyfriend went up my shirt last night, and I totally felt his erection through his jeans!"

"Get out, really?"

"For reals! I was like 'ew gross' and then like 'oh my God.' And my mom caught me talking on the phone with him at midnight!"

"Oh no! Did she know it was him?"

"No, I said it was you—she's so stupid, she had, like, no idea."

When our band teacher separated us, Christy passed me notes about how her boyfriend wanted to have sex, but all she wanted to do was make out. She liked to make out; it was one of her favorite things. As proof, Christy wore a necklace full of soda can tabs.

"Each tab is a promise that your boyfriend will give you eight hours of Frenching. *Hard* Frenching."

Christy wore about five hundred of these tabs, given to her, I imagine, by either an incredibly thirsty boyfriend or a generous Coca-Cola distributor (also probably a boyfriend). Not only is that a whole lot of soda, 6,000 fluid ounces to be exact, but also a whole lot of Frenching—4,000 hours. Not bad for a twelve-year-old. I have to admit that even at my age now, I've only accomplished a fraction of that. I mean at 4,000 hours, you'd have to make out for at least ten hours a day for a year. Now, I'm not saying it can't be done, but it's virtually impossible for anyone who has a job or goes to school or breathes.

In seventh grade, Christy ruled the dance floor—she could do the Roger Rabbit, the Kid 'N Play, and even the Robocop. She went into each dance move seamlessly, sometimes throwing in a little Cabbage Patch to dazzle onlookers. But her most famous move was a dance she invented in which she paraded around the room jutting out her massive chest, sticking out her ass, and wagging her round hips back and forth. At first it didn't drive the twelve-year-old boys crazy—they were still figuring out that they were supposed to be turned on—but eventually they started staring and asking her to do the Dance during lunch. Christy happily obliged; she thought of the Dance as her mating call. She was exploring her sexuality when many girls, including me, were arguing over which flavor of Jolly Rancher was best (cherry). Christy showcased what was to come: boobs and boys.

Quickly the girls in junior high caught up with Christy and by the end of seventh grade, most of the girls were wearing soda can tabs—not a single soda can within a four-mile radius had a tab. Soon Tanya was totally into Brian who was into Kara who was going with Jason who kissed Tanya (but it was by accident so he wasn't cheating). All the girls got boobs and asses, and they loved being women—except, of course, for me. Friends looked to Christy for advice on love, for she was an experienced woman.

One afternoon Christy broke down in tears. Her mom had forbid her to French Jim, a boyfriend she had just met at the Fallbrook Mall (one who had just turned seventeen). All of our friends offered our sympathies and advice, but since I had never even talked to a boy, much less Frenched one real hard, all I could do was hairspray her bangs and make her a mix tape. One afternoon, in a solemn ceremony, Christy offered me a soda can tab, explaining that she had snuck out of the house and finally Frenched Jim for eight hours at a super-secret location (probably his truck), and it would mean a lot to her if I wore it as a sign of our sisterhood. I was touched. I wore it until my mother asked why I was wearing trash and made me take it off. But I kept it in my locker so I could wear it at school. I secretly hoped that everyone would assume I had a marathon make-out session with a sixteen-year-old.

All the seventh-grade girls seemed to get their first periods around the same time. Girls in the locker rooms and bathrooms would ask each other for a tampon or a pad, and then launch into a heated tampon-versus-pad discussion, which segued into conversations about boys—Do you really think he likes me, as in like-like me? When it came to feminine products, I'd smile and pretend I knew what I was talking about. I was a staunch pad supporter because I didn't know how to use a tampon, even though Christy explained it to me several times and even drew diagrams, which I squirreled away in my nightstand for when I'd need them. I figured I'd support the feminine product that mystified me less.

"Oh my God, I totally use Always Ultra with Wings, too. The Dri-Weave like totally works."

Everything I knew about being a woman, I learned from Christy or commercials. Christy knew I hadn't gotten my first period, but she didn't care. I don't think anyone in school actually cared, except for me. Christy helped me create the illusion of womanhood by talking periods with me and I was grateful. I believed that menstruation was a requirement to fit in, just like MC Hammer pants and high-top sneakers.

The rungs of my junior high's social ladder were made of feminine products. Just by asking Laura Paris if you could have a tampon, you established three things:

1. You were a woman.
2. You knew that Laura was a tampon supporter.
3. You were personal enough with Laura to ask
   for a tampon.

If Laura presented you with a tampon, you established three things:

1. You were a woman.
2. You were like totally cool.
3. She would one day ask you for a tampon,
   thereby forging a relationship between two
   mature women in which you could do womanly
   things together, like drink coffee and talk about
   how annoying men are.

Sometimes I'd ask a fellow woman for a pad, just to gain some menstrual cred. Then I'd stow it in my locker, hoping that one day I'd use it. I was worried. What if they found out I wasn't a woman at all? What if I never became a woman? What if I had to pretend to feel bloated and get cramps for the rest of my life? How would I get out of gym class?

Because my ovaries were in a deep coma and possibly dead, I was short, skinny, and flat in eighth grade. If I turned to the side, I'd disappear almost completely. The boys began snapping bra straps, and girls would giggle in response, happy that they were getting attention from the opposite sex, even if it was childish and perhaps a little painful. I was paralyzed with fear. I had no bra to snap. I began wearing white undershirts and folding them up so that they looked like sports bras. Perhaps I could fool them. The day Jon Thomas grasped for my back, he took a handful of tank top. He thought he had missed. I was relieved to tears.

When we reached ninth grade, Christy and I went to separate high schools. I would have to face womanhood without my woman. Every single girl in high school wore a bra and they all needed one—even the small, skinny girls (who were still significantly taller and heavier than me). In the locker room, women openly mentioned their unmentionables, oh did you get a new bra? Yeah totally I got it from Victoria's Secret. Oh my God it is so cute! Before and after track-and-field practice, I made sure that no one would discover that I didn't wear a bra and didn't even need one. I would change in the bathroom stalls or wait until women had left the locker room. I was late to practice most of the time and as a result, my coach made me run extra laps around the track. Still, jogging an extra mile didn't take nearly as much energy as pretending to be a woman.

But soon I became desperate. I was fourteen years old, with no hair, rack, or rear in sight. Anyone could confuse my back for my chest. I had the body of a nine-year-old boy. It was time for serious measures.

I had always pictured a moment where I would walk with my mother around a lake and tell her how I had become a woman. Everything would be beautiful and soft. There'd be a lot of ducks. We'd be wearing flowing white dresses and she'd give me a hug and a box of Tampax.

Since our family didn't live near a lake and I hadn't gotten my period, my fantasy of connecting with my mother over plastic applicators dissipated. I sighed and walked into the kitchen. She was rinsing rice, and I watched her pour out the white, cloudy water into the sink.

"Mom, I'm not a woman yet."

"What?"

"I haven't gotten my period yet."

"Anne, you worry too much. Why you want period?"

"Everyone else has it. I should have it, too."

"I tell you, when you get period, you not want it."

"But I want it. And I don't have it. There must be something wrong with me. Like I'm missing a tube or I have no eggs or something. Maybe I should see a doctor."

"You know Mommy got period late so you get late too. Anne, you normal, only crazy. Maybe you can see acupuncture."

"No, no needles, I can't do it. OK, fine, can I at least buy a bra?"

"Why you want bra? You have no breast!"

I must've looked wounded because an hour later we arrived at the Promenade Mall. We walked into Robinson's "intimates" department, where I spied an eleven-year-old woman picking bras off the B-cup rack. She held up a white lacy one and showed her mother, who seemed to approve. I wished the entire store were empty so I could be alone in my embarrassment; there were hundreds of bras in every shape and color and I was sure none of them would fit me.

"Excuse me! Excuse me! Hello!"

I whipped my head around and saw my mother shuffling up to a store clerk.

"Yes, how may I help you?"

"Where I can find bra—very small? For my daughter . . . Anne, where you are? Anne!"

I ducked down behind a rack of nightgowns and robes and hoped the entire world would go boom. Nuclear fallout was a more merciful way to die than embarrassment.

"Jockey makes a nice set of training bras. All cotton and very plain."

"Oh good, good, my daughter like plain. But has to be small."

I buried my face in a robe and wondered if I could hang myself with the sash.

"Yes, we have them and they're on sale. You can look over there."

I poked my head out from the rack and saw the clerk point. I followed the direction of her finger and saw gigantic red sign that said "PRICE BLOWOUT! AAA-AA CUPS ON SALE! EXTRA 10% TAKEN OFF AT REGISTER." All it needed was sirens and a flashing neon arrow. I gasped and my mother spotted me and dragged me to the rack. I looked around nervously, hoping I wouldn't see anyone from school. She held up a pink cotton bra.

"How about this?"

"OK-sure-looks-good-I'll-try-it-on-bye!" I grabbed the bra and sprinted to the dressing room.

My mother passed me a hundred bras underneath the dressing room door and I refused to show her how they fit. Because they didn't. I cursed my ovaries and then pleaded with them: *I hate you, you're ruining my life, no wait, I don't mean that, please wake up, please don't do this to me.*

"Anne, you take too long. How about this?" She passed over another bra.

"No it doesn't fit. Give me the size down."

"OK Mommy go look."

In the dressing room mirror, I stared at the baggy, white bra. The elastic around my chest was loose and the straps kept sliding off my shoulders. There was no underwire and lace, unlike the bras my mother or girls at school wore.

"Anne, it smallest. Maybe I ask lady if they have more in back."

I winced. I pictured my mother talking to the clerk. They'd talk about how inadequate my boobs were and laugh. Then the lady would go into the backroom, which was filled to the ceiling with a million bras, and discover that they didn't carry bras that small, but that maybe I should check out Toys "R" Us because she heard Playskool made My First Bra.

"No, no, no, Mom, it's OK. This one fits."

I grabbed one, hoping that one day I could fill it out, and threw it at my mother. I escaped to the cosmetics counter and sniffed perfumes while my mother paid at the counter.

My first bra was made of white cotton and because the straps were too long, I had to tie knots in them so they wouldn't slip off my narrow shoulders. The clasp was in front and it rubbed against the skin on my sternum, making me bleed. As uncomfortable as my bra was, it was still a bra. It made me feel like half of a woman, which was better than being just a girl. My bra's cups, which were really more like tablespoons, remained unfilled for the next three years.

Puberty finally hit me at Karen Crocker's house. We had just started our last semester of high school and we were studying together. I went to the bathroom and discovered dark spots on my underwear. At first I was suspicious, as if somehow I was wearing someone else's underwear. Then, after I realized that the underwear and the blood were both actually mine, I felt surprised. I'd imagined that I'd know when I was about to get my period—that for four to five days before I'd retain water, have painful cramps, get easily fatigued and irritable, and feel tenderness in the breasts (which I didn't have)—that was what the Midol commercials had promised me. But I felt pretty good that day, not even a backache. Then, my surprise turned quickly into annoyance. I had ruined a perfectly good pair of underpants and even though I had been waiting for my period for the last four or five years, I was completely unprepared. Still sitting on the toilet, I reached over to open the medicine cabinet and the drawers under the sink. Nothing. I groaned. I yelled for Karen through the bathroom door.

"What's up? Is there no toilet paper? Did you check under the sink?"

"No, I need something else."

"What?"

"You know, something *else*." I heard her laugh through the bathroom door.

"Come on, Karen."

"OK, it's not a big deal, which kind do you need?"

"A pad."

A minute later I heard a knock on the door.

"OK I got one. Open the door."

"I can't open the door. I'm on the toilet." I heard a stifled laugh. Then I saw a small pink package being shoved underneath the door.

I had expected to feel overwhelming relief after getting my first period, but I didn't. If anything I felt a little embarrassed, but for the most part, I felt exactly the same as before. Except now I was wearing something that felt like a diaper. So much for being a woman.

I drove home later that evening and walked into my mother's bedroom. She was watching the Korean news.

"Hi, how you are? You hungry?" She said this to the TV, where a very stern man was delivering some very stern news.

"I got my period. Cool, right?"

Her eyes remained locked on the TV.

"Mom, pay attention, I got my period."

Nothing.

"MOM!"

"OK, OK I just kidding Anne! It joke! Of course it cool. You worry so much, but Mommy knew you get soon. But now you think 'Why I want period? I get cramp and fat and I feel terrible.'" She puffed out her cheeks and rubbed her belly. I laughed; it was the same motions she used to make fun of my overweight brother.

"But it's a good thing."

"Good, yes, tomorrow Mommy go to Costco and buy pad. Maybe I get chocolate, too." She smiled, turned back to the TV, and waved me away. I guess this was the closest to flowing white dresses and ducks we would get, which was fine with me.

The next day I told my coach I couldn't run because I had "cramps," which was not only a code word for bleeding, but also a lie many girls used to get out of practice. He looked wary.

"Really? You've never skipped practice because of cramps."

"But they've never been this bad before."

I went back to the locker room and packed up my gym bag. A teammate asked why I was leaving. I explained that I was riding the crimson wave and my Aunt Flo had come to town, and wasn't that just a bloody shame?

HOLY Crap

"Anne? Come here, help Mommy!"

"I'm in the bathroom!"

"NO, come now!"

"I said, I'm in the bathroom. Let me just finish."

"What you do in bathroom?"

"I'm throwing a big party. What do you think I'm doing?"

"You *hurry*! Come quick!"

I threw aside my book and grumbled. My mother has an uncanny ability to interrupt my father, brother, and me during our

most private moments. I can imagine her waiting until we get settled on the toilet before bellowing orders through the bathroom
door: Take out the trash. Wash the car. Put the dishes away.
And no matter how mundane the task, it always sounds like an
emergency, as if the fate of modern civilization rests on my sepa-
rating the darks from the whites. I must admit, it's a very clever
strategy—we're literally caught with our pants down and willing
to say yes to anything just to stop her hollering through the bathroom door.

"*Ayoo*, what take so long?"

"Can you just hold on for a minute?"

"Why you not help Mommy? Why you not love you mommy?
How I raise such daughter?"

I flushed the toilet, quickly washed my hands, and scurried into
the kitchen.

"You spend so much time in bathroom! Maybe I move you bed
there!" She laughed and I rolled my eyes. My mother pointed to
the wet spots on my jeans where I had dried my hands. She sighed.
"Why you not use towel?"

"Because you told me to hurry! What do you want? What's
up?" I tapped my foot impatiently.

My mother pointed to a box neatly wrapped in brown butcher
paper. It was large and flat—the kind of box used to ship framed
paintings. We only had one painting hanging in the house and I
think it was there when we moved in. It was probably a mirror.
My mother liked mirrors, and we certainly didn't need another
one.

"Help me open."

"You got me out of the bathroom to help you open this? You
couldn't wait for another five minutes?"

"Just help you mommy. It surprise."

"Surprise for me?"

I raised one eyebrow suspiciously. She had given me a surprise a few weeks earlier—a set of SAT preparation books and a weighty pack of vocabulary flash cards. I had explained rather irately that a car was a more suitable surprise for a fifteen-year-old.

"No, this surprise for everyone."

Together we ripped the brown paper off and opened the box. Inside there was an unruly mound of bubble wrap, and I immediately pulled some out and marched on top of it. The snap of firecrackers echoed off the kitchen walls.

"ANNE!"

"Oh come on, it's fun, try it." I threw some at her feet, which she kicked aside.

"Why you always bother Mommy?"

I spied a corner of a wooden picture frame inside the box. "What is it?"

"A picture."

"Well, duh, but of what?"

"Someone special."

"Of me?"

My mother scoffed and we lifted the frame out of the box. The picture was rolled up in white bandages like a mummy. I started unwinding it carefully. Maybe it was a picture of our family, though we never took a group portrait. Maybe it was a painting by a family friend. Finally I just ripped the bandages off.

I gasped. I stared. And the face of Pope John Paul II stared back at me.

"OH. MY. GOD."

The Pope wore gleaming white floor-length robes with a short cape tied around his shoulders. A gold chain with a cross shone on his chest, and covering the top of his balding head was a small white disk of cloth. Facing the camera, he stretched out his arms with palms turned up, as if he were waiting for a hug. The Pope seemed ghostly, a pale figure standing in front of a dark and empty background. His sunken eyes gazed steadily ahead and he looked eerily peaceful and solemn, as if he were waiting for good news that he already knew about. But the most noticeable characteristic about the Pope was the size. The picture was three feet wide and over three feet tall—a massive, monolithic, full-body shot of the pontiff that floated in a sea of bubble wrap on our kitchen floor.

"Oh no, what have you done? Where did you get this?"

"From church."

"You have to give it back. I can't believe this. It's so big. I mean, look how big this is." I stood next to it so my mother could see that the top of the picture came up to my waist.

"Anne, I tell you, it not big, you short."

"No way. It's like a hundred feet tall."

"It POPE. Be nice."

"But why does it have to be so *big*?"

"Shh, Anne, shhhh."

"What, you think he hears me?"

The glossy Pope was framed with a matte border that, upon closer inspection, was not matte at all, but white silk with a faint floral pattern. The frame itself was dark, polished wood, about five inches thick, and a plate of heavy glass protected the photograph. It was a nice frame, of solid construction. Immediately I thought of other pictures that could replace John Paul II, maybe Kurt Cobain or even a pleasant landscape. I examined the back and sadly discovered that the photograph was sealed inside the frame, a transparent vault that captured one man in one pose for the rest of my life. My mother pulled out a tape measure and noted the dimensions. She looked up from the photograph and scanned the empty walls in the kitchen.

"NOT EVEN. You CANNOT put this thing up. It's horrible. And it's heavy. It's gonna tear the wall down."

"No, no, we can put up."

"But it's gonna scare everyone."

"Anne, stop."

"Jesus Christ, just look at it!"

"*Anne!* You mouth!"

My mother's eyes burned right through me. She walked out of the kitchen and down the hallway, scanning walls and assessing lighting options. I followed closely behind. She stopped and gazed at an empty space between two windows.

"No, you can't put it there, there's not enough space."

She brushed me aside and pulled out a tape measure. Luckily, I was right. The picture was too wide.

"How about in the closet?"

"Anne, stop, I get very mad."

"Ok, fine, how about behind the door in the closet?"

She ignored me and walked into the den. She looked curiously at a large photograph of Mike and me when we were little; we were standing in front of a waterfall on Jeju, an island off the Korean peninsula. I was wearing a white ruffled blouse with puffy sleeves and pants so pink they looked like I had gone wading in a pool of Pepto-Bismol. My brother's green collared shirt hung on for dear life around his protruding belly and his striped tube socks were pulled up to his pudgy knees. We were both grinning at my father behind the camera; I had small white nubs for front teeth.

"Wait, you're going to take down that picture for HIM? You don't even KNOW him."

"Anne, I think you talk too much."

"You are not allowed to touch that picture."

"Anne, go away."

"MOM, YOU CAN'T PUT IT THERE!"

"OK. OK, *ayoo.*"

She growled and walked toward the front door. She stood pensively in front of a large, empty wall.

"Oh no way. Not here."

"No, it good, Mommy like here. Shh, why you not be quiet, Anne?"

"But it's too close to the front door!"

"So?"

"So? So people will SEE IT."

She reached for her tape measure. I plastered myself against the wall and stretched out my arms so she couldn't measure.

"Anne, you move, NOW."

"No, this isn't right! I can't let you do this. You're gonna have to kill me first."

My mother stormed away and yelled for my father. He promptly showed up with a toolbox and a stepstool. My heart sank but I stood steadfast against the wall.

"Dad, please don't do this to us."

"Annie, you have to move."

"Why don't you put it in your bedroom?"

"Because Mommy want put picture here."

"I'm sure it'd be OK if you hung it in your bedroom."

"I don't want it in the bedroom."

"Well I don't want it next to the front door."

"Annie, move out of the way. I'm sorry, but you lose."

"We all lose."

He peeled me off the wall, set down the step stool, and plugged in his drill. Using museum-grade wall anchors and four-inch nails, my father hung up the picture. I stared at him in disbelief; he was aiding and abetting poor taste, a sin really.

"But don't you think it's a little too big, like maybe just a little over the top?"

My father sighed and looked down at me from the step stool. "I think it fine. Be nice to you mommy. You both yell and scream, I get such headache, you know? I'm an old man and you make me older."

A few hours later I heard the front door open, followed by a gurgling, choking noise and a deep-throated laugh. My brother had discovered the pale-faced stranger staring at him. I joined Mike in the foyer.

"Dude, is this some kind of joke?"

"I know, I know." I shook my head.

"What the hell? It's so *huge*. She's totally lost her shit."

We stood in front of the photograph, dumbfounded and oddly absorbed. The picture had a peculiar magnetism to it, like a piece of eye-torturing art or pornography. I wondered what kind of shoes John Paul II was wearing under those robes. What footwear possibly goes with papal garb? Dainty, soft-soled slippers? Italian leather dress shoes?

"Dude, why did you let her do this next to the front door?"

"How is this my fault? I tried to stop her. And, you know, it could be worse. It could be in your bedroom."

He shuddered. "It doesn't even go with the house."

The picture hung near a long scroll of traditional Korean calligraphy and a blue vase painted with a scene from a fifteenth-century Korean countryside.

"Nothing goes with this. Except a church."

"Well, can we take it down?"

We weighed the possibilities: my mother's wrath (and, perhaps, God's) for deposing the Pope or the constant abuse from friends for this jumbo JP2.

"I doubt it. Dad anchored it to the house. It's never coming down."

My friends' reactions, upon seeing the Pope greet them at the door like a restaurant host, ranged from poorly stifled surprise ("WHOA, WHAT THE HELL IS THAT?") to complete shock (dropped jaws, followed by awkward silence). One friend with a highly developed sense of irony was impressed ("Holy shit, where can I buy one?"). Since most of my mother's friends were from her church, they fawned over the photograph, complimenting its beautiful frame and saying what a wonderful Pope he was and oh don't you know he speaks eight languages and likes experimental theater and poetry and thank God he wasn't assassinated, what a tragedy that could've been. My mother seemed quite pleased with herself and the oversized display of her faith.

For a long time, church was mostly a social outlet for my parents, especially my mother. They opened up their house for bible study groups and fund-raising meetings, during which they gossiped about other church members and their children (drugs, marriage, Harvard) instead of discussing the bible. My mother sang in the choir along with her three closest friends and chaired committees that organized picnics and holiday parties. When my brother and I were in elementary school, our parents dragged us to church even though the services were done in Korean, which we tried our best not to understand. My mother's elbows kept jabbing us in the ribs to keep us awake in the pews and after Mass, we waited impatiently in the car for our parents to stop chatting with their friends. We honked the horn and blasted the radio while our parents tried to ignore us. Eventually the church added services in English, but Mike and I remained uninterested. Who cares about turning water into wine? At thirteen and ten years old, we were too old for magic; we wanted Nintendo. Mike and I usually left in the middle of Mass to go across the street to eat at Tommy's Hamburgers. But even seasoned french fries weren't enough motivation to leave the house on Sundays. Every Sunday morning my brother and I fought a two-front war; I would refuse to get into my church clothes and Mike would take sanctuary in the bathroom. Eventually our parents gave up on us. We'd have to find salvation on our own.

Before the Pope came to live in our house, we had a cross here and there, but the photograph took religion in our home to a whole new level. It showed me that, to my parents, being Catholic was no longer just about socializing and finding a community of Korean immigrants, but about having a relationship with God. As a fifteen-year-old, I wasn't pleased about this. Being religious was uncool—I knew this from watching the Church Lady on *Saturday Night Live*—and I had my own opinions about abortion and contraception. I thought my parents had been doing just fine as laissez-faire Catholics, but now they were announcing their beliefs to the world, or at least to anyone who visited us. We didn't need organized religion and we didn't need John Paul II upsetting what was once a pleasant-looking foyer. But the picture was just the beginning.

Gold crosses started to appear on the kitchen and living room walls, and colorful rosaries emerged on every lamp in the house. Jesus candles and dishes painted with biblical scenes materialized on the mantle. A wooden manger, complete with lambs and a baby Jesus, became the centerpiece of the living room table. Little statuettes of Jesus (on the cross and off) and pewter figurines of saints convened on sidetables. A picture of The Last Supper on petrified wood hung near the bathroom, not in the kitchen or dining room where one might have supper. A two-foot-tall ceramic Virgin Mary sprung up in the garden; she looked serene among the flowers and the jackrabbits that ate them. On top of the piano, my mother placed a framed prayer and a bust of Jesus, which I always turned toward the wall when I messed with the timer so I'd only practice for fifty minutes instead of an hour as my mother had instructed. Korean Catholic books, with covers picturing clouds and rays of sunshine or praying hands, sat next to her romance novels on the shelves, and they were all squeezed between matching Jesus and Virgin Mary bookends. An enormous three-foot-long rosary, with beads the size of baseballs, hung on the wall over the television. Our house was getting out of hand. There was nothing that could be done short of lighting it on fire.

"Mom, this is enough. You're even ruining the bathroom."

She had placed a framed copy of the poem "Footprints" between the soap dish and the toothbrush holder in my bathroom. The poem is about a man who reflects back on his life and sees two sets of footprints, a set that belongs to him and a set that belongs to Jesus. The man discovers that during the hardest parts of his life, he sees only one set of footprints. He gets angry because he thinks Jesus abandoned him. It turns out he was wrong: "During your times of trial and suffering, when you see only one set of footprints, it was then that I carried you." That was in my bathroom, the least inspirational place in our house. "Ruin? Why ruin? I think nice."

"Then put it in your bathroom."

Etched on a mirror in blue type, the poem was not enlightening, but just one more thing I had to clean. It often got stained stubbornly with dried lather and toothpaste spit.

I'm not sure why my mother collected all this kitsch. Why does anyone collect anything? You buy one thing, it makes you happy. So you buy a hundred more. I see it all the time—my friend's mother collects ceramic pigs, which may or may not be worse than my mother's religious paraphernalia. It's possible that my mother thought having more Catholic knickknacks made her more Catholic, although I have trouble believing that she attached deep spiritual meaning to a plastic Virgin Mary golf tee carrying case. All I know is that by the time I was seventeen, our house looked like a roadside Catholic gift shop.

During my senior year of high school, my mother and members from her church went on a pilgrimage to Medjugorje, Bosnia-Herzegovina where the Virgin Mary allegedly appears and sends messages through six chosen people. It's a site not officially recognized by the Catholic Church, but that really didn't matter to my mother. She brought back a staggering number of statuettes. She unwrapped each one carefully and lined them up proudly on the kitchen counter. By this point I should've been accustomed to all my mother's holy trinkets, but these statuettes both depressed and angered me. All the bric-a-brac littered around the house were eyesores and embarrassments, but visiting a war-torn country to see a ghost defied logic. She was in the same category of people who worshipped vegetables, mud stains, and cow pies that look like the baby Jesus.

"What's this?" I pointed to an angel with wings outspread. He was wielding a menacing spear and stepping on the head of a man, who looked quite uncomfortable. In contrast, the angel looked very tranquil.

"It angel, so nice you think?"

"Why is he stepping on that guy's head? Angels aren't supposed to do that."

"Because, that man was bad man."

"What did he do?"

"Bad thing."

I rolled my eyes and moved on to a wooden cross with a gold Jesus. "You bought another cross? We have like a million of them."

"This one from church in Bosnia."

"But it looks exactly like the one you got from downtown L.A."
I turned over the cross. "It says 'Made in China.'"

"Really?"

"No."

She smacked me on the shoulder. The cross ended up above the enormous photograph of the pontiff, which had once stood out in the Korean-ness of our home. Now it acted as the focal point that tied the house and all its Catholic kitsch together.

During the 1994 Northridge earthquake, most of my mother's figurines broke, and the floor was strewn with pieces of Jesus, some of which cut into my bare feet when I ran to take cover. No one in the family was hurt and our house suffered no structural damage (just hairline fractures in the drywall and cracked floor tiles), but my mother agonized over her keepsakes. Many of the larger statuettes fell on top of the more delicate ones, pulverizing them into a powder. The Pope, however, clung resolutely to the wall. Not even a bit off center. The Pope survived a 6.7 shake in his usual mild manner, much to my disappointment. My mother tried to repair some of her figurines with glue, but it was difficult to figure out which bleeding heart and which praying hands belonged to which figurine. Unfortunately, the earthquake only gave her an excuse to buy more Catholic tchotchkes and within a year, my mother replaced all of her keepsakes and added more, including a plastic angel that doubled as a toothpick holder and a cross with a coat hook.

A few years ago my parents decided to downsize to a smaller house, and I returned to Los Angeles to help them pack. The first thing I did was turn my attention to the kitsch.

"Where are you gonna put all this crap?"

"You mouth so dirty, Anne."

"The new house is smaller. There's no room for any of this."
I held up a box of laminated Jesus and Mary trading cards with
prayers printed on the back.

"I find room."

"Mom, seriously, you have to throw this stuff out. Give it
away."

"No, no, I don't think so."

"You won't even know it's MISSING."

I paused. I recalled all the stuffed animals my mother had doled
out to those snotty Korean kids and my unwillingness to let them
go, to surrender my beloved one by one. I remembered the time
spent loving each one and the pants-wetting joy of getting another.
And I could see how my mother thought my animals were multi-
plying beyond control and her concern over my obsession. I could
see why she wanted to thin my herd.

But what came to mind wasn't sympathy. It was payback.

I gleefully removed several plastic rosaries hanging from the
lamps in the living room. Years of dangling near blazing light bulbs
had melted the beads into blobs.

"I'm throwing these away."

"No you can't, Anne."

"They're melted. You have a thousand more." I showed her the
blobs and flung them into the trash. My mother winced. I held up
a set of glass Jesus candles. "These are going, too. They're done
burning anyway." I tossed them and they clinked together in the
trashcan, like champagne glasses toasting. My mother reached into
the trashcan and I swatted her hands away.

"ANNE!"

"Look, there's no more wax in them. You can always buy more.

I promise I'll buy you another one." I smirked. I looked around hungrily and grabbed a chipped plate depicting one of the Stations of the Cross.

"NO, ANNE, NO! SAVE!"

"OK, OK you can keep this one." I set it aside. I picked up a clear plastic Jesus with a gold wire halo around his head. "What is this?" I threw it away. My mother opened her mouth to protest. I cut her off. "Listen, the new house is half the size of this one. Where are you going to put all of this?"

My mother shrugged. I had a point. Score one for Annie. With each dusty and chipped statuette I threw away, I thought of the stuffed animals my mother had given away when I was little. I was doing this for the koala in the Dodgers jersey, the white seal, the kitten. I relished the thud of the figurines in the trashcan and when they actually shattered or broke, I tried to hide my delight. Moving had never been this fun.

"No, Anne. Stop. Put in box and I save in garage. Don't put all in trash! You understand?"

"Some of this has to be tossed." I showed her a Precious Moments figurine of a praying child. Her leg was fractured—an earthquake injury probably.

"IN BOX, ANNE!"

I rolled my eyes and grabbed a box. I started adding my mother's keepsakes: a few Jesus figurines, plastic saints and angels, Virgin Mary magnets, prints of nativity scenes, candles. One box led to another, much bigger box. Bookmarks with inspirational prayers, a chintzy Saint Christopher paperweight, a knicked plastic music box that played a flat hymn. When my mother wasn't looking I deftly carried a few boxes to the trunk of my car. Three of them went to Aardvark's Odd Ark, so that they could wind up in someone else's house, and one box of items in disrepair went to my favorite charity, the trashcan.

But there was one more item. After fifteen years, it was time to detach JP2 from the wall, from my mother.

"There's no way you are bringing this."

"Anne, it Pope, yes I bring."

"Where are you going to put it? There's no room in the new house."

"I make room."

"You're being totally unreasonable. There's no place to put this. No wall space. Look how old this looks. We have tons of photographs to bring over. *Of us.*"

Little spots of mildew stained the white silk border. Fifteen years worth of dust sat on top of the frame, just out of reach from my mother's rag. I found a gray moth that had crawled behind the picture to die.

"I have to bring. How I can throw like trash?"

"Well how about you give it away. I'm sure your friends wouldn't mind taking in the Pope."

"No, no one want."

She was right. Who would want an old, oversized Pope? Not one of her friends; they probably had one in their houses anyway. It would have to be one of my friends.

"OK what if I told you my friend wanted it."

"Who?"

"J.D. He's Catholic. He loves the Pope."

J.D. is Catholic and he does love the Pope. He has a Pope-Soap-on-a-Rope in his shower. He has a Pope-O-Scope, a kaleidoscope of the pontiff he fashioned out of a cardboard tube. He had decorated his house with Jesus candles and in his bathroom had installed a bleeding heart nightlight that cast a creepy red glow over the toilet. The first time he saw JP2 gracing my mother's wall, he was nearly moved to tears of ecstasy. I didn't even bother asking J.D. If he didn't want it, the trash would gladly take it.

"He want picture?"

"Yes, he wants it."

"You not lie?"

"Why would I lie? Look, I'm telling you, there'll be no room for this."

"I don't know." My mother sighed.

"You can't bring it all. Where you gonna put the picture of The Last Supper?"

"I don't know."

"So, J.D. has the perfect place for the Pope." I wasn't sure where this perfect place was in J.D.'s house, but as far as I was concerned, any place that wasn't in my parents' house was perfect.

I loaded JP2 into my car and drove it over to J.D. He greeted me in his driveway and nearly fainted from excitement when I showed him the pontiff lounging in my backseat.

"Are you sure your mom doesn't want this? Like seriously?"

"Trust me. It's all yours."

"This is the coolest thing ever. Really. I love it all hard." He grinned and carried it into his house.

J.D. wanted to put the Pope in his bathroom but couldn't find enough wall space. His girlfriend refused to let the John Paul II into the bedroom. His housemates weren't excited about having a Pope in the living room. Or in the kitchen. They cursed me for passing on the picture, as if I had passed on gonorrhea. J.D. wandered around his house surveying the walls.

"Don't sweat it, J.D. I mean, if you can't find a place for it, you can throw it away."

"Throw it away? No, no, even if I wanted to, I couldn't throw it away."

J.D. wrapped the Pope in an old, filthy blanket and shoved it in the dark corner of his garage, between boxes of dusty magazines and outdated computer parts. To the best my knowledge, it's still there today.

When my parents finally settled into their new home, there were noticeably fewer figurines gracing the tables. My mother managed to exercise some self-control. She bought a cabinet where she displayed many of her Catholic treasures alongside her china. There were, however, countless boxes of religious decorations tidily stacked in the garage, and I regretted not throwing out more boxes when I had the opportunity.

Still, each time I visit my parents' house, I notice more and more crosses and pictures and candles and statuettes. Slowly, my mother has been adding to her collection: tissue box covers embroidered with crosses, a wine decanter etched with prayers, a clock depicting Jesus' last days on the cross (something J.D. would kill for). I hope that they'll move again, and I wonder if praying for a minor earthquake makes me a horrible person.

THE BEST DIET

Despite all the pressure to get into Harvard and achieve my parents' American dream, I ended up at the University of California at Berkeley, which may not be the best school in the country, but it is the best public school, at least according to *Newsweek*. At first my parents grumbled—maybe I should've taken advanced placement physics or maybe I should've done more volunteer work or cured diseases in Africa. Eventually they awoke from their crimson haze and learned to love the Golden Bear.

My first week of college, I got a phone call from my mother every half hour. Mornings were a particularly rough time—she's one of those criminally insane morning people who believe that if they are up at dawn, the rest of the world is, too. On my first day of school, my mother called at 6:00 and asked why I wasn't getting ready for school, and I groggily explained I had class at noon. At 6:30, she wondered why I still wasn't up yet. At 7:00, she asked what I was going to eat for breakfast and I replied, "sleep." At 7:30, she announced she was going to go for a quick walk, so she might not be home if I called her. At 8:00 she told me how nice her walk was and that I should walk every morning, too. At 8:30, she called in a panic, that if I didn't wake up NOW, I'd miss my noon class. I growled that I didn't need three and a half hours to get ready for class, but I did need three and a half hours of more sleep. At 9:30, she inquired about my plans for lunch. When the phone rang at 10:00, my exhausted roommate blurted out, "Please, please tell her to stop!" I explained to my mother that she didn't have to wake my roommate and me up every morning because we had a special machine called an alarm clock that served the same purpose. But the phone calls still didn't stop. She called to tell me she dropped off her dry cleaning. An hour later, she told me she bought a new brand of hair gel. When she called to tell me to save all my mismatched socks because she had found a few loners in her laundry room, I threatened to cut her off forever.

I understood that she was lonely, now that her children were out of the house, and I understood that she worried about me, especially as the baby in the family. But the calls were out of control, beyond what was reasonable and healthy. I assured my mother that I was responsible—I was going to all my classes on time and even found a job—and that I didn't want or need to know that her grocery store reorganized the fruit section. Finally, my mother agreed to talk once a week, and my roommate and I got an answering machine and a phone with a ringer we could turn off.

My first semester went smoothly; I enjoyed independence and life without my mother. I realized how much she held me back—from irresponsible boozing, bouncing from party to party, and smoking an acre of weed, all on a school night. I can't say that I missed her too much, but I did like talking to her once a week. One March afternoon, I called to chat with my mother, and I was surprised when my father picked up the phone. I normally called him at his lab, where he spent his fourteen-hour days analyzing metals and compounds and watching the Lakers or Dodgers on a tiny black-and-white TV.

"Hey, Dad, why you home? Shouldn't you be at work?"

"I'm taking the day off."

"Cool, what for?"

"To take care of you mommy."

"Oh did she sprain her ankle again? Or throw her back out? She's always doing that kind of stuff."

Once on a ski trip my mother swerved to miss a tree and collided into my father instead. He was fine, but she sprained her knee and had to be taken down the mountain by the Snow Patrol. She was really embarrassed because she fancied herself a good skier. Whenever we waited in line for a chair lift, she adjusted the zippers on her puffy orange snowsuit, tightened her enormous amber goggles, and practiced going into a low-tuck position. She hopped from left to right and pretended to race down moguls, making a *shh shh* sound through her teeth.

"No, no, Annie, not this time."

"She got a cold or something? Just tell her to take some NyQuil and she'll get over it. You should take some too; she's probably contagious."

"No, not a cold."

"Then what? What's left? You give her a rash or something?" I laughed at my own joke.

"Mommy has cancer."

I lost control. I felt as if my body had lost its shape, as if my muscles and bones had melted away. My hand loosened around the receiver and I dropped it. I slowly slid off my chair and slumped to the floor, my head gently resting on the carpet. She has what? For some reason, all I could think about was my dorm room carpet. It reeked like feet and beer and I wondered when the last time it had been shampooed, certainly not in my lifetime. This carpet probably had lice. Maybe scabies—can scabies live in carpet? My feeble brain was unable to process *Mommy* and *cancer* in the same sentence, so it moved on to something it could deal with—filthy carpet infested with mites. Somehow, I summoned the strength to pick myself up and climb back into my chair. I took the receiver and lifted it to my ear, but the spiral cord got tangled into an enormous dreadlock and left about two inches of slack. I tried to untangle it, but got frustrated, so I leaned over to talk into the receiver with my nose an inch from the rest of the phone.

"Annie, you there? Hello? You OK?"

"I don't know."

"She has breast cancer but it OK, I think we catch it in time. Don't worry, Annie, nothing to worry. She got biopsy. Then she got surgery; she got mastectomy. They took it out, the whole thing."

"Is she going to die?"

The question gushed out of my mouth. The words were so ugly and grotesque; I wanted to swallow them again. What a horrible thing to say. Why did I say that? My father was silent. No, I thought, don't say it again. Don't do it.

"Is she going to die?" I felt nauseous and I had a funny taste in my mouth, something sweet and sour. I remember reading in a magazine that some people tasted metal in their mouths before they got heart attacks. I wiggled my tongue in my mouth. No, not metal, Fruity Pebbles maybe.

"I tell you don't worry. She start chemotherapy already."

"When did this happen?"

"Three month ago."

Three months ago? My hand, which had gone limp just moments before, now gripped the phone so firmly I could hear the cheap plastic crack. She had been diagnosed three months ago, in January—probably right after I started my spring semester. We had talked so many times since then, and she seemed fine. She had gossiped about her friend's tragic plastic surgery and complained about my uncle's unruly six-year-old who was already expressing interest in explosives. She asked me if Berkeley was safe at night and if I had classes with any hippies. But I remember once she sounded sick and tired and she blamed it on the flu. I guess it was cancer. I never once thought anything was wrong. How could I have known if they didn't tell me?

I thought about a moment in sixth grade when I came home from school to find my mother packing her china into cardboard boxes. When I asked what she was doing, she mentioned nonchalantly that we were moving in two months and asked why I was so surprised because the family had been discussing it for a long time. It was the first I had ever heard of it. Our house wasn't even for sale—they were going to sell it after we moved to the new house, which they already bought without showing my brother and me. I asked Mike about the move and he responded, "We're fucking moving?" My parents never told us because they knew we would get angry. They waited until the last possible moment, when the bad news could no longer be hidden and they needed us to pack up the house.

"Dad, you knew about this three months ago and you didn't tell me? She had a goddamn *mastectomy* three months ago and I didn't know about it?" Heat flooded into my chest and I started shaking from fear and fury.

"Annie, we not want to worry you."

"Worry me? You didn't want to fucking worry me? Why didn't you tell me? I've been talking to you this entire fucking time and no one told me shit. Why would you fucking do that to me?"

I slammed the receiver down. No, I thought, Don't be like this. I immediately dialed again.

"Dad, I'm sorry, I'm just . . ."

"It's OK Annie. Everyone be OK."

"I'm coming home."

"No. You can't come home. You have to finish school."

"What the hell are you talking about? School's not important."

"No, Annie, no. Mommy and me want you to finish school. You have month and half more. You finish first."

"No way, I'm coming home now. I can't finish the semester. How can I finish school? Tell me how I can fucking finish school. Do you think breast cancer will be on my finals? Because that's all I'll be thinking about."

"Annie, don't make this hard on Mommy and me. Mike is around, he help."

"Let me talk to Mom."

"Annie, you help by staying in Berkeley. If you here, you will give Mommy stress because she worry about you, because you worry about her. You understand? I going to hang up now. Mommy sleeping."

I was angry. I was angry she had cancer and I was angry that I couldn't go home to help; they didn't need my help. Families are supposed to rely on each other and they had cut me off. I felt so betrayed and useless. Mike went to school two hours away from my parents' home, so he found out earlier. Everyone had known except for me. I called my brother.

"What the fuck, Mike? Why didn't you fucking tell me?"

"Annie, I couldn't. You know that I couldn't."

"But you're my brother. You're supposed to tell me shit like this. That's part of the whole brother-sister deal."

"What could you have done? No one can do anything except for the doctors, don't you get it?"

"Mike, is she OK? Tell me the truth, is she OK? No one's telling me anything. It's all fucked up."

"She's tired all the time; she sleeps a lot. She lost some weight. Look, she's going to be OK."

"Promise?"

"Annie, come on. Don't make me do that. Don't be a pussy. We'll take it slowly, dude."

I sat in my room for the rest of the afternoon and I cried. Theresa, a round-faced girl who lived on my floor, dropped by to see if my roommate was home. I tried to explain between my sobs, but only key words would come out—"Mom . . . cancer . . . Chemo . . ." Theresa's brown eyes welled up with tears.

"Hey now, you got to keep it together."

She tucked me into bed and sat for a while, stroking my hair. People were coming back from classes and parties were beginning. The sounds of clinking bottles and gangster rap and inebriated voices and high-fiving leaked into my room, so Theresa put on music to drown out the noise. I woke up the next morning, to find that my roommate had left me alone for the evening and the

CD was still playing—Theresa had put it on repeat. It was the soundtrack to "Little Women," the campy version starring Winona Ryder, Susan Sarandon, and a lot of bonnets. With my eyes swollen shut from weeping, I reached for the phone.

"Mom?"

"Anne, you OK?"

"I want to know if you're OK."

"Oh, I fine, don't worry. You know I just tired. I sleep a lot and chemo make me throw up, but not too bad. I watch TV, I read. But I get so much headache. All time headache."

I was surprised by her even voice. She sounded a little tired, but strong. No wonder I couldn't sense anything was wrong for the past three months—she was good at hiding any fear or pain. Before I called her, I was worried that I was going to lose control and cry into the phone. But if she could keep it together, I could too.

"That's not good. What are they from? Did you tell the doctor?"

"I get headache because everyone worry and I say, why you worry? I not dead. But everyone worry. So I tell my doctor and he say there no cure for worry." She laughed at her joke. I forced a dry chuckle.

"We care about you, that's why we worry."

"You know what make Mommy very happy?"

"What?"

"Daddy stop smoking. I so happy, he smell much better now."

I thought about what would happen if my father had cancer, too. No, I thought, stop that. "Can I come home to see you?"

"Anne, you make easy on Mommy and Daddy, OK? You stay in Berkeley, get all A and then go Harvard."

"Mom, you know I love you, right?"

"I know, Anne, I know. Everyone know."

"Because I don't know what I'd do without you."

"Oh Anne, you sound so . . . so . . . what word you always say when you see bad movie on TV. Like movie about man who drink too much. Then he get divorce and life so sad. His life so much trouble and everybody cry and you get headache?"

"*Cheesy?*"

"Yes, yes, you sound very cheese. Promise me you not worry."

"Promise me you'll tell me everything."

Just as my mother had pestered me with incessant phone calls when I first started school, I began running up my phone bills. I called every hour to find out how her chemotherapy was going, what she was eating, what medications she was taking, or what she was watching on TV. She explained that therapy was going slowly and her hair was falling out, but she didn't mind so much. ("I have so much hair before, you can't tell it get thin.") My grandmother made her favorite meals, especially baked mackerel and bean sprout soup. She took all her medications, whatever was on her nightstand. She forgot what all of them did; she just knew when to take them. She started watching American soap operas but found them hard to follow. I offered to watch them and explain each episode, but she didn't like the idea because it interfered with my studying.

A few weeks passed and just as she was coping with the nausea and fatigue from chemo, she developed swelling and an infection in the scar tissue where surgeons had removed her breast. She went to the hospital to get the area drained and then she had an allergic reaction to the anesthetic. Within minutes she ran a fever and developed bright red hives all over her body. I called her at the hospital.

"Oh I so itch, Annie. Itch everywhere. The itchy worse than cancer. Even on my face! All I want is scratch everything. You should have seen Mommy. Like mosquito bite all over. And I so hot. I sweat like crazy person. My pajama get so wet. But doctor gave me medicine and I feel better now."

"Well I'm glad someone feels better."

The doctors instructed her to stay in the hospital for two nights, though she wanted to recover at home. She wanted to read in her own bed with the comfortable king-sized mattress and down pillows, not the twin-sized adjustable mattress with bars on the side. The hospital's reading selection, she explained to me, was boring. She preferred her trashy Korean novels over the self-help cancer survival books or healthy lifestyle magazines. She wanted my grandmother's comforting stone-pot stews, not the institution-alized meals that came divided into four sections on a plate. ("Why hospital food have so much potato?") My mother hated spending more time at the hospital than she had to—she already visited the doctor every week to receive chemo drips and check-ups. Still, the rest of my family liked having her in the hospital because the around-the-clock patient care gave us a sense of security, a feeling we rarely felt.

After my mother's infection was under control, her sister-in-law picked her up to drive her home. As my mother and aunt were driving through an intersection just a few miles away from my parents' home, a driver ran through a red light and crashed into the side of their mini-van. My aunt was shaken up, but fine. My mother, however, was not as lucky. The impact and the seat belt damaged her chest tissue, which was still recovering from the infection. An ambulance picked her up and she was readmitted into the hospital. She needed surgery again.

"This is so fucking ridiculous, Mom. Why is this happening to us?"

"Anne, you stop cry. You mouth so dirty, I get bleach."

This time, she sounded weak and fragile. She wheezed from the pressure of the bandages. She talked softly and slowly and her mouth was dry. She couldn't drink anything twenty-four hours before going into the operating room. It was the worst I had ever heard her, and I thought that maybe she wouldn't make it. Cancer, an infection, an allergic reaction, a car accident, a daughter with a dirty mouth. It was as if someone was trying to get rid of her. I wanted to be there.

"Can I at least come home for the weekend?"

"Anne, shh, it OK. You promise me you not worry. Please stay at school. You have no reason to come."

"No reason? You're the reason why I'd come. Just to see you, don't you want to see me?"

My mother was silent. I tried not to get angry; this wasn't allowed to be about me.

"Anne, sweetheart, please. You come home for summer vacation in three week. You can wait. I don't want you see me like this. On outside I look worse than I feel on inside, you understand?"

I heard her voice shake, just a little. Or maybe I thought I heard it shake. I couldn't remember the last time I heard my mother cry. Crying was not something she did; she was too tough for that. When my mother told her mother about the breast cancer, my grandmother started tearing. My mother scoffed, "Don't be such a baby, I'm fine." As far as I know, nothing, not even cancer, could make her cry. At least not in front of people. She was careful to keep up an illusion of strength and for the most part it worked, or maybe everyone let her think it was working.

"It's just not fair, why did he have to crash into you? Why couldn't he hit someone else? Anyone else? Why did he have to hit the one with cancer?"

"No problem, you know? I do surgery before, very easy. I go sleep and doctor fix and I wake up. So easy for me. Nothing to worry about. How school? You meet any Korean boy?"

"Mom, this isn't funny. Cancer's not funny."

"You know what funny?"

"What?"

"Today Daddy tell me I got jury duty. I got jury duty, how funny! I think jury duty worse than cancer. Maybe I tell them I have cancer and they feel sorry for me, you think?"

For my mother, I wrote a request for an excusal from civic duty and sent it to the Jury Commissioner's Office, citing extreme physical impairment. I attached a copy of her medical records, highlighting the long list of prescribed drugs that stopped her cells from dividing so quickly—drugs with side effects that could impair judgment in a courtroom. I also attached a copy of the accident report, in case breast cancer wasn't a good enough of an excuse. My request was granted.

The surgery to repair her damaged chest tissue went well, much to everyone's relief, and my mother recovered a few days in the hospital. My father told me she handled it well, just like her mastectomy and chemo.

"You mommy very tough. I think is she tougher than me."

"You're tough, too. We're all tough."

"Yes, I know. But you mommy, I don't know how she does it. She's like a machine. Like the Terminator. Nothing can stop her."

My father picked her up from the hospital and drove home very slowly, taking side streets and avoiding busy intersections.

I had two weeks left in the semester, and I managed to finish somehow. I received mostly B's, with a C+ in Introduction to Anthropology, the lowest grade I had ever received in any class. I decided not to tell my mother. There was no sense in pissing off a cancer patient—I didn't want her losing any more of her hair worrying about my grades. I packed up my dorm room and left Berkeley without securing housing for next year. I figured I would sort it all out if or when I returned next fall. Part of me thought that I wouldn't return and would remain in Los Angeles to help my family and watch my mother recover. But the other part of me knew that my mother would rather die than see me take a leave from college. She wanted me to continue with my regular life, not stopping or even slowing down for her cancer. I think I understood this, but it was frustrating nonetheless. I flew home.

My father picked me up from the airport. He seemed so old; my mother's cancer had aged him significantly. His hair was more gray than black and his skin seemed too baggy for his bones. His eyelids drooped from worry and exhaustion, as did the skin around his mouth and chin. He gave me a hug and helped me with the luggage. For the first time ever, I noticed he didn't smell like cigarettes and instead smelled like sweat. I wasn't sure which was better.

"Mommy's doing OK. She doing chemo again and doctor say she doing good. A lot happen in last five, six month. Everything go so quick."

I realized that I had never asked how they discovered her cancer. On the car ride home, my father told me how my mother started getting tired easily and had no energy to do her daily activities, like going to the grocery store or cooking.

"She didn't even have energy to go shopping! So I think she must be sick!"

At first they thought she had the flu, with mild aches and pains and fatigue. She took a lot of Tylenol, and when she didn't get better, she went to see a friend from church who was a doctor. The doctor said it was probably a virus that was going around and it needed to run its course—nothing serious. Then one afternoon she took a nap and didn't wake up, even when my father came home. Whenever she's home, my mother always greets my father in the kitchen when he returns from the office. This time, however, she remained in bed and when my father found her, he tried to wake her. He wanted to know about dinner—he is incapable on his own in the kitchen—and she just wouldn't wake up. My father took her to the hospital where they ran tests and found a tumor in her breast. It was the size of a tennis ball.

"It was as big as a tennis ball? That's insane."

I find it weird to hear stories about patients with tumors the size of softballs or baseballs or golf balls—how does that all fit in there? How do you not notice a sporting good lodged in your breast?

"We couldn't believe it. We were so surprised. You think flu, and really it cancer. But you Mommy was so calm. She said, 'OK, what I do now? How we can we make this better?' Very business, but I know she was worried. She wanted surgery right away."

My father talked about her mastectomy and the chemo and how doctors discovered she had a mild heart murmur, which complicated surgery.

"She has a heart murmur, too? Christ."

"It very common. She had it her entire life and she never knew. She has a valve that work a little slow—a lazy valve. The doctors have to be careful in surgery and with certain medicine, that's all. Nothing serious."

My father seemed so calm and matter-of-fact. He talked about my mother's cancer with the composure of someone who had hundreds of conversations about it. My parents have a lot of friends and relatives; I'm sure he had to update all of them several times a day. For him, cancer had become part of his life like a routine. But I knew from his aged face that it hadn't been easy for him.

"Thanks, Dad."

"Oh, Annie, it OK. I don't mind picking you up from the airport. My office is very close."

When I arrived home, my mother greeted me in the kitchen. I was shocked. For the past few months, I had envisioned her as bald and thin, with degenerating muscles barely clinging to her frame. I imagined that her eyes, not her body, exuded the spark I heard in her voice most of the time. But to my surprise, she looked terrific. Her hair was thinner, but still considerably bushy. I could see a mass of gray roots—she hadn't dyed it in awhile. Her skin was still soft and smooth; she has always been adept at applying make-up and made sure to moisturize every night. When I kissed her, I discovered her cheeks were still as soft as they were when I was six years old. She wore a baggy, pink flowered dress that disguised her chest bandages. I loosely wrapped my arms around her, not wanting to disturb what swelling and scarring that lay beneath her clothes. The only noticeable change was in her movements. She walked slowly and avoided gesturing her arms wildly, something she always did to add emphasis to her words.

"Wow, Mom, you look really good. I mean really, it's amazing."

"Anne, you so silly. I always look good."

"But no, you look *really* good."

"You know why? Because I lose weight."

My mother is not petite. She has an athletic build, with broad shoulders, strong arms, and long legs. She is voluptuous with wide hips and generous curves, but she never needed to lose weight. It wasn't the inches she lost on her waistline that made her look great. Clearly, she had the look of a woman who was beating cancer—a triumphant glow. I wondered if cancer even had a chance against her. If there was a moment when she was scared of dying and leaving her family and friends behind, I didn't know about it. She would never tell me anyway.

"You just don't look like all the cancer patients on TV. They are really skinny and wear big scarves wrapped around their bald heads. You're not bald."

"No, Anne, I lose weight, I like it. I think cancer is best diet."

"Mom, that's horrible!"

"I tell everybody at church, get cancer so you can look like model." She strutted a few steps, with her hands at her hips.

That summer, I spent most of my days taking care of general household duties. I did laundry and dusted and went grocery shopping. For my mother, I rented her Korean movies, brought her tea and pills, and shuttled her to the hospital. Whenever she got home from chemo, she'd shuffle straight into bed and sleep quietly. After a few hours, I'd tiptoe into her bedroom and hold my breath so I could listen for hers.

My most exhausting duty, however, was not taking care of my mother—I was happy to help in that area. It was dealing with the kitchen. My grandmother visited us two days each week and cooked meals for the entire week. When my grandmother cooks, she brings out every single plate, utensil, and pot from the dark recesses of the cabinets. She litters vegetables all over the floor, leaves fish entrails and bones on the counter, and splashes soy sauce and sesame oil all over the stove. I spent a lot of time scrubbing food and grease off every surface in the kitchen, including the walls. As I cleaned, I thought about how my family could've used my help while I was away at college—no one can scour a kitchen sink quite like me. I was hurt that my mother didn't want me to see her weak and vulnerable. I guess because I was the youngest and my mother and I were close, she wanted to protect me. She thought I couldn't handle her cancer, or maybe she was afraid that I could handle it—that I was growing up.

"Mom, you know what the worst part about your cancer is? It's the kitchen. How does Grandma make such a mess?"

"I know, I think she crazy. She cook so much and I not so hungry."

"But you have to eat—it's important."

"My medicine make me so dizzy and I throw up."

I looked at her, wishing there was something I could do. Well, there was one thing.

"I heard from somewhere—maybe I read it or something I don't remember—that marijuana helps with chemo. It helps you get your appetite back and stuff like that."

She raised her eyebrows. We had never talked about drugs; it was understood that drugs were illegal and not something anyone should do. I went through D.A.R.E. in fifth grade, but I also lived in Berkeley.

"Is that right? But mari-wan illegal. Can *you* get mari-wan, Anne?"

I stopped. I sensed an ambush. Even though my mother was weakened with cancer, she was still sharp and capable of entrapment. There was no way my mother would smoke weed. I couldn't imagine rolling a joint for her or teaching her how to use a bong. She didn't even drink alcohol—she told me she always hated the taste and found no enjoyment in the effects. Was she trying to incriminate me or was I just being paranoid? She knew Berkeley's reputation of "experimentation," but she had no reason to believe that I had been experimenting. Why did I even bring this up?

"I don't know where to get marijuana. It's probably hard to get since it's illegal. But if you're interested, I can try to find out. There might be someone from Berkeley who'd know."

"No, no I don't want it. Drug very bad for you. I hope you never do it."

My mother keeps her breast on her cosmetic table. Among the bottles of anti-wrinkle cream and toner and palettes of eye shadow, there is a gelatinous flesh-colored mound of silicon. The prosthetic doesn't feel like a real breast—it's much squishier and has no nipple—but it mimics the weight and shape of one. It sits in a special bra that has a soft cup for her healthy breast on one side and a special pocket for her prosthetic breast on the other side. She always wears her bra over a thin tank top because the elastic chafes the sensitive skin of her scars.

The right side of her chest, where her breast used to be, there is a wall covered in pale, soft skin. Small mounds of white scar tissue speckle the area. Underneath the thin, mottled skin, there is a layer of strong chest muscles that stretch over her chest plate, which protects her heart with a lazy valve. When the supportive bandages first came off around her chest, eight months after her mastectomy and three months after her car accident, her right shoulder kept on rising to meet her ear in an awkward half-shrug. There was no breast to weigh her shoulder down and no bandages to hold the shoulder muscles back. The muscles in her shoulder and chest had not adjusted yet. She kept on using her hand to push down her shoulder.

"It won't go down. It make Mommy so frustrate."

Under the guidance of her doctor, she learned exercises to help loosen the muscles in her shoulder and stretch and strengthen the muscles in her chest. The exercises were painful, I could tell by her wincing when she practiced lifting her arms straight to the side like an airplane.

"How badly does it hurt? Maybe you need to take a rest."

"It hurt but you know, I have to do. I have no choice. I have to practice so I can play golf."

Even after she completed chemotherapy and went into remission and her scars had healed, my mother did not want reconstructive surgery and a breast implant. She wanted that part of her body and that part of her life gone forever.

"Why I need breast? I have no baby, I not need breast, right?"

"I thought I was the baby!"

"Anne, you have your own breast."

So her doctor fitted her with a prosthetic breast. She wears the same clothes she has always worn; she never wore low-cut blouses or dresses so the prosthetic is never a problem. She plays golf better than ever, placing in tournaments. She's in great health; in fact, she's probably healthier than me. When I visit her, she gives me hard hugs and crushes me against her chest, and I forget about everything.

Occasionally, when my mother is lying in bed, under her electric blanket watching TV or reading, she calls for me. She doesn't bellow my name as she normally uses, but instead she whimpers. "Anne . . . Anne . . . you there?"

I immediately stop what I'm doing and run to her side. My mother does not tolerate weakness, so when she whimpers and groans softly in pain, I start thinking about tumors again. Maybe this time it isn't a tennis ball, but an orange or a grapefruit growing in her body somewhere. My stomach tightens and my fists clench and my brain struggles to shut off.

"Anne?"

"I'm here, are you OK? You look pale."

"Can you get your old, sick mother a glass of water?" she says in Korean. She smiles at me weakly.

"Are you OK? Do you want me to get you anything else?"

"No, no I'm OK. Maybe I need hot tea."

"Just tea, anything else?"

"Maybe some chocolate."

"OK, tea and chocolate. That's it?"

"Cookie." She laughs softly.

"OK, tea, chocolate, cookie."

"And fruit." She laughs a little louder. "Don't forget bring knife. And napkin. Maybe you can go get ice cream for me? I like green tea mochi from Trader Joe's." She explodes into laughter and claps her hands together. "I'm too lazy to get myself."

It's a cruel joke and I fall for it every time, but I never laugh harder.

vegetarian
enough

In my sophomore year high school, one of my best friends read *Diet for a New America* by John Robbins and decided to become vegetarian. Eating animals, Alyson explained to me, was bad for the Earth, bad for your health, and like totally bad for animals. Livestock was pumped full of antibiotics, hormones, appetite stimulants, and tranquilizers and then they were debeaked, dehorned, and castrated so they could wind up on our kitchen table and in our bodies where their flesh would slowly fester and poison us and cause heart disease, tumors, and a black soul. I did not want a black soul; I wanted to keep it fresh and yellow. Like a squash. So, I became vegetarian too. I didn't even bother reading the book. I figured if John Robbins could convince Alyson, then he could convince me. Done and done.

"WHAT?"

"No meat. I'm vegetarian." My mother and I were sitting on the living room floor, folding laundry. I crumpled Mike's shirt into a tight, wrinkled ball and tossed it aside. I figured if he wasn't going to help us, then I wasn't going to help him. "It's a better way of life."

"Better for who? You not eat meat, you get very sick. Then you die."

"Actually, you're wrong. Meat makes people sick. It's bad for you."

"What you mean bad for you?"

"It causes heart attacks and stuff."

"No, Anne, you cause heart attack. Who tell you this?"

"John Robbins."

"Who John Robbin? You friend at school? Teacher? I want have talk with him."

"No, no, his family started Baskin-Robbins."

"He ice cream man? What do ice cream man know? When you get so crazy? How can you be Korean without meat?"

"Grandma is Korean, and she's Buddhist and vegetarian." Actually I knew my mother's mother wasn't really vegetarian, but I thought I'd try to slip that by my mother.

"Anne, Grandma eat meat!"

My grandmother is quite active at her small Korean Buddhist temple in Los Angeles. I used to accompany her to her temple, which is actually a two-story house converted into a temple, and play with my stuffed animals behind a large golden statue of the Buddha. I also used to take a burning stick of pine incense and run around with it, pretending to be an Olympic torchbearer.

"Yeah I know, I know, but she's not supposed to eat meat. She's gonna go to Buddhist hell, where everyone is vegetarian." I laughed at my own joke. Clever, clever girl.

"So that mean you go hell too because everyone in heaven eat meat."

OK, maybe not that clever. Mother 1, Annie 0. "Whatever."

"Grandma break rule because she know meat is good. She only vegetarian at temple. She say she vegetarian enough." My mother reached over and unfolded one of my shirts I had folded. "Fold this again. Why you fold like monkey?"

"It doesn't matter—I'm still not eating meat. And I can fold my own shirts anyway I want."

"I not cook for you. You starve."

"I can cook for myself."

My mother laughed. "Like what?"

"I can make spaghetti."

"Everyday? Who eat spaghetti everyday?"

"Italians."

"Who you are? Sophia Loren?"

"I can get a cookbook. I can make us all vegetarian food. We can all be healthy together."

"No one eat it."

"More for me then."

My mother sighed and studied my face. Her eyes slowly moved over my eyes, nose, and cheeks. She seemed to be looking at me for the very first time; I was growing up and making my own bad decisions about my life, just like normal adults. "How long you be vegetarian? One year?"

"Forever."

"Forever?"

"Forever."

"That long time, Anne. Forever is until you die."

"I know what forever means, and I mean it. Forever."

"You eat fish then."

"No, fish is meat, and I don't eat meat."

My mother smiled slyly as she folded a pair of my pants. "You remember when you little you want to be mailman. You want to drive little car. Then you want to be ballet dancer and you take one ballet class and you say, 'Mommy, my feet hurt so much,' and you cry *waah waah*."

"This is it. I'm a vegetarian. I'm never eating meat again."

"OK, we see."

The next week my mother prepared all my favorite dishes, which all happen to have meat: chicken stewed with potatoes and carrots, brisket slow-cooked in soy sauce, kim chee stew with pork, soy bean stew with clams and shrimp. I knew what she was doing, trying to lure me back to the dark side, the side rich with protein and iron and low in carbohydrates, but I remained strong. This was not a war over meat, but like any teenage rebellion, it was a war over will, Annie and John Robbins vs. Mother and Pretty Much Everyone in the World, including America and Korea. Luckily, Korean food is pretty vegetarian-friendly. Each meal has several side dishes that consist of some kind of salted, pickled, or fried vegetable. I was perfectly content to forego the main dish, which once stood proudly on four legs or had wings.

"What's wrong with Annie?" My father looked at my mother, confused. "She's not eating. Is she sick?"

"Very sick. Ask your daughter."

"What's wrong with you?"

"Nothing."

"Then why you not eat?"

"Because I don't eat meat."

"When did this happen?"

"Like forever ago. Last week. Where have you been?"

"I've been eating meat. Why are you vegetarian?"

"I already went through this. Meat is murder. It's bad for everyone. Even Earth." I wrapped a piece of dried seaweed around some rice and stuffed it in my mouth.

"You have to eat meat. How will you live without meat?"

"I've been living pretty awesome without meat."

My father looked to my mother for reinforcement. She rolled her eyes. "Don't bother," she said in Korean, "your daughter does what she wants to do."

"You have to cook her something without meat."

"I'm not cooking anything special for her. She says she can take care of herself."

"She's lying. She's fifteen, what does she know?"

"Actually, I'm sixteen."

"Eat meat."

"No."

"I told you not to bother. Just leave her alone."

"I mean it Annie, eat meat."

"No."

"How could you let her do this?"

"I didn't let her do anything. She did this all on her own."

"Dad, it's not even an issue. It's just the way it is. Besides, I'm totally full and didn't even eat meat. How about that?"

Eyeing my bowl of plant matter, my father cringed. "Why do you do this to yourself? How will you eat my steakie?"

My father calls steak "steakie," which is a derivative of the Korean word for *steak, suh-tay-kuh*. My father prides himself on two things in the kitchen. The first is "steakie" for which he takes two slabs of London broil and sprinkles salt, pepper, garlic powder, and MSG on each side. After he's done broiling it, he uses scissors to cut the meat into perfect one-inch cubes that are tough and chewy enough to unhinge jaws. His second specialty is Rice Krispies treats, in which he uses margarine instead of butter, cuts the amount of marshmallow, and adds peanut butter, vanilla extract, peanuts, and raisins. The mixture is extremely crunchy, with the texture and flavor of drywall with peanuts and raisins. He packs it into a pan so tightly that they turn into sandy bricks that shred the roof of my mouth into a bloody mess. Once my father caught me making normal light and fluffy Rice Krispies treats, using the traditional, unaltered recipe that the good people of Kellogg's developed, tested, and approved, and he looked so hurt that I never did it again. I used to plead for both of my father's specialties until I tried the real versions, but by then it was too late. My father had built his entire identity in the kitchen around chewy meat dice and drywall.

"I guess steak is the one thing I'll miss." I smiled apologetically and helped myself to more sautéed bean sprouts.

Because I like to make my life as difficult as possible, I became vegetarian right before Thanksgiving. My family always hosts all the relatives for Thanksgiving and my mother spends the day cooking a traditional turkey dinner plus a large Korean meal. I told her once many years ago that Thanksgiving was too much work, that she should just stick to the turkey and skip the Korean food, and she scoffed and asked what kind of meal didn't have Korean food? When I explained that the pilgrims probably didn't eat Korean food, she laughed and said that they missed out—had they eaten Korean food they probably wouldn't have starved.

During the course of Thanksgiving dinner, my brother outed me and announced to the entire family that I had become vegetarian "like some kind of dirty hippie" and soon all my aunts, uncles, and cousins were riding the express train to Nagsville.

"You're going to get jaundice, and if you're lucky, you'll die."

"Leave me alone, Mike."

"Do you eat anything that casts a shadow?" My cousin Andy is athletic and health-conscious, but even he decided that eating vegetarians was better than being one.

"Hey, isn't your belt made of leather?" My cousin Woo-jay asked me in Korean, "Isn't that made from animals?"

"No, it's fake," I lied.

"What about seaweed, do you eat that?" my uncle asked me.

"Yes." I sighed and concentrated on my broccoli.

"What about yogurt, do you eat that?"

"Yes."

"What about eggs, do you eat that?"

I groaned. "Yes, Uncle, yes."

"Why not become a vegan? A Communist one. Start some kind of revolution against meat. Against everything delicious with lungs."

"I said, leave me alone, Mike."

"But you're just skin and bones, if you don't eat meat, you'll start fainting," my aunt remarked in Korean. She tried to put turkey on my plate but I moved it away.

"Please, no turkey. Really, I'm fine."

"OK try some of this," my uncle slopped a spoonful of stuffing on my plate.

"There's turkey in it!" I pushed the stuffing to the side.

"But you can't see the meat so it doesn't count. Stop being difficult."

"Meat tastes good, you should eat it," Tina said simply. She is the nice one in the family. "Just eat some turkey and everyone will leave you alone."

"Everyone, please, I'm fine, there's plenty of food here." The truth was there was a ton of vegetarian food: broccoli casserole, mashed potatoes, yams, beets, fried rice, three kinds of kim chee, scallion pancakes, fried tofu, and fruit salad. Clearly, there was enough food to feed an entire kibbutz of dirty vegetarian hippie revolutionaries prone to fainting, but without meat, I would starve to death, if jaundice didn't kill me first.

After my mother got cancer and then beat it into submission, she became more conscious of her health. She bought an enormous juicer that was powerful enough to squeeze juice from a rock, and proceeded to throw in everything she could get her hands on: apples, oranges, carrots, tomatoes, celery, beets, lettuce, persimmons, more or less everything in the produce aisle. She started making juices from strange combinations like apple-tomato-beet and celery-spinach-orange and offered me some, which I declined. I explained that I preferred to chew my vegetables. She talked to both her Western and Eastern doctors about dietary supplements and started taking vitamins, something I've never had the energy to research. She started eating a lot of oatmeal because it was better for her stomach, which was prone to ulcers, and drastically reduced her caffeine intake, another thing I've never been able to do but know I should.

"Anne, see? Now I eat like you. Vegetarian."

I had returned to college to start my second year and knew my mother had changed her eating habits, but I would never have thought my mother would actually become vegetarian. "What? Seriously? You don't eat any meat at all?

"No, no meat. Only I eat fish, a lot of fish."

"That's not vegetarian. You eat fish."

"But I eat a lot of vegetable. So healthy! I think my skin feel better. How come you skin so dry? Maybe you not eat enough vegetable."

"I just have dry skin."

"Eat more vegetable."

"I do eat vegetables. That's all I eat. I'm vegetarian. A real one."

"Then maybe you need to eat fish."

I groaned. I can never win. "So you're not eating beef?" There is always some form of beef on the table.

"Doctor tell me I have to watch my cholesterol, so I eat very, very little beef. Better for me. See? I'm vegetarian, like you."

My mother gave me a few tofu recipes that were simple enough to prepare on my own and talked about the fruits and vegetables that were high in antioxidants and vitamins that could improve my skin. She told me to drink a lot of water and eat more beets. I imagined she sounded a lot like John Robbins, though I didn't really know since I hadn't read his books.

A few weeks later, my mother had fallen off the quasi-vegetarian wagon just as fast as she had gotten on it. She decided that life with very little mammal and poultry was not a life she wanted to live.

"Vegetarian so hard. What you eat?"

"I eat plenty."

"I get so hungry, I keep eat and eat and eat. I waste so much time."

"I thought you liked to eat. Everyone likes to eat. You know, people complain when they don't eat."

"Everything taste same."

"No they don't. Spinach and mushrooms do not taste the same, and tofu tastes different too."

"I miss beef and chicken and pork too much. They miss you too."

"That's ridiculous. They miss me because they know I won't eat them."

"Eat meat."

"No. Be vegetarian."

"No."

"I guess we're back where we started."

Sometimes I think that my mother and I are very close to finding a common ground, to finding an area of interest that we can explore together and bond over and discuss without bickering or nagging each other. But then I realize this is impossible, that if we didn't give each other a hard time, we probably wouldn't get along.

I don't remember what chicken tastes like, nor do I recall the flavor or texture of pork and beef. I imagine them to be rubbery and grainy and maybe a little squishy. As of this moment, I've been vegetarian for thirteen years, four months, twelve days, and nine hours, plus or minus thirty minutes. I'm approaching the moment in my life when the number of years spent as an omnivore and as a vegetarian are equal, and yet, every time I talk to my mother, she asks me what I ate for dinner and if I'm still vegetarian, as if I'm still going through some kind of a phase. She'll leave messages on my voicemail: "Hi Anne, it Mommy. You still vegetarian? I get worry because you not eat meat. You have to eat meat again. Ok, it warm in L.A. Bye, bye." Why does she do this? Does she think that I'll suddenly stop and think, yes, you were right all along, please give me some beef, and then my mind will be blown and I'll start listening to everything she says? Could beef be the gateway drug to pork, chicken, and ultimately, obedience?

To be honest, I'm not certain that I ever truly cared for the vegetarian cause. Sure, it could be a healthier way of living, better for the environment and all of its creatures, but I'd be a big, fat liar with flaming, leather pants if I said that becoming vegetarian wasn't about rebellion and making decisions about my own life. As a teenager, vegetarianism fulfilled both these needs. Plus, I thought it was kind of cool. It made me feel unique, set me apart from other high-school kids—albeit in an inconvenient way—and annoyed my parents, and what adolescent wouldn't want that? I'm sure my parents know that the choice I made when I was sixteen was not one I actually thought through, and my mother is laying in wait for the day when she can point and say, hah, it was all just a silly phase, one that lasted over thirteen years. But today, my being vegetarian is more about inertia than principle. I haven't eaten meat for so long, it's no longer part of my vocabulary. I'm set in my dietary ways, just the way meat-eaters are, and even if I wanted to give up being vegetarian, I couldn't. There's too much at steakie.

THE DEVIL MOISTURIZES

In my grandmother's opinion, everything was too something—the weather was too cold, food was too spicy, clothes were too itchy, the neighborhood was too loud and too dirty. Even fruit was too sweet. My father's mother complained about everyone: people on the street (rude), store clerks (fraudulent), housekeepers (lazy), her children (thoughtless), and her grandchildren (disrespectful, reckless, and spoiled). She never approved of the women her four sons married, and she constantly reminded her daughters-in-law of their lower status (you come from a poor family) or lower intelligence (you never went to college). In short, my grandmother was a bitch. I never liked my grandmother, and I'm pretty sure my relatives never liked her either. They're just too polite to admit that she is a miserable woman. Still, we all understood that she was old—she wouldn't and couldn't change—and so we silently endured her. After my grandfather passed away, she became the eldest in our family, the matriarch of the Choi clan. Her position demanded respect and we showed it to her. It was the Korean way.

When I was in elementary school, my family and I visited my grandmother and our Korean relatives every year, but as we got older, the trips became less frequent. By the time I was in high school, most of our relatives had moved to the States. The ones that remained in Korea, with the exception of my grandmother, preferred to visit us. Seoul, they explained, didn't have Las Vegas or the Grand Canyon or bagels. By the time my mother and I took a trip to Seoul in the winter of 1998, I hadn't been in about ten years. I was stunned by how much the city had changed. There was more of everything—people, cars, skyscrapers, factories, subway stations, lost tourists. My favorite department store, the one with an entire floor dedicated to stuffed animals and dolls, was now nestled between a Kentucky Fried Chicken and an All-American Burger, which is Korea's fast-food chain that competes with McDonald's. I gawked at a three-story Haagen-Dazs packed with young Korean hipsters and wondered how they could eat ice cream in the middle of December. Then I wondered how they could even eat ice cream. I guess Koreans and dairy have an abusive relationship. On the congested streets, billboards and jumbo televisions competed for my attention: a wafer-thin Samsung cell phone with a built-in mp3 player, camera, and PDA; the faster and more luxurious Chairman sedan from Ssangyong; the always refreshing and always tasty OB Lager; Amore White, a cosmetic designed to soften and whiten skin. Posters of Tom Hanks and Meg Ryan were plastered on the side of every bus, urging Koreans to watch them fall in love in yet another movie, this one called *Yoobu Gotu May-eel* (*You've Got Mail*).

Though in ten years Seoul had changed from a modest city to a modern one, my grandmother's house was as exactly as I remembered it. According to my mother, my grandmother's house hadn't changed in fifty years, maybe more. The same battered furniture my father and his five siblings grew up with was still in the house, in exactly the same places. The table with the loose leg still leaned against a corner; the wooden chairs still had the same ripped cushions, and my grandfather's armchair, threadbare at the armrests, still stood eerily in the corner. During the last years of his life, my grandfather enjoyed looking around the living room, enchanted by details that everyone else overlooked because they weren't stricken with Alzheimer's. On the shelves there were picture frames that held white pieces of paper with ghostly outlines of landscapes and people; the photographs had faded in the sunlight decades ago. There was a still rip in the rice-paper screen door. When I was six, I had accidentally kicked a hole in it and my grandmother spanked me. No one other than my parents had ever punished me and when my grandmother took her open hand to my bottom, I pleaded with stinging eyes to my mother, who turned her head and told me that I should listen to my grandmother.

My grandmother refused to install indoor plumbing in her house even though she could afford it. She probably had the last three-bedroom house in Korea without modern conveniences. She preferred to urinate in an oversized brass teakettle; the way Koreans did in a bygone era. When I was younger, she would order me to empty it at inappropriate moments, for example, right before dessert. I think it was a test of obedience, which I always passed. I'd get up from the table and carry the kettle outside and dump it in the empty alley, which reeked like stale urine and hot trash. Then my grandmother would order my mother to pour me a glass of weak barley tea, which looks a lot like urine.

The only thing more intolerable than my grandmother's personality was her smell. Ever since I was little, a thick odor came from her mouth. It was as if her tongue were decomposing, and I imagined a mass of flesh-eating maggots in the back of her throat. Her gums were lined with a few brown nubs that posed as teeth but served no real purpose. Her wrinkled mouth always had mysterious crumbs in the corners, though there were no cookies or crackers in sight. Whenever I had to hug her, the stink overpowered me and I could smell it on my own clothes hours later. My grandmother's smell hung heavily in the house, seeping into the furniture and rugs.

When my mother and I walked into my grandmother's house, the odor singed my nose; her stench was exactly as bad as I remembered it. Though I hadn't seen my grandmother in a decade, she wasn't exactly excited to see me, which didn't surprise me, and I wasn't exactly excited to see her, which didn't surprise her. My mother and I greeted and bowed to her and she ordered us to sit down for lunch. I glanced at my mother nervously. My mother had reminded me that it was important to eat everything that was served, despite the putrid quality of my grandmother's food— everything she cooked actually tasted like vomit. My mother and I had argued over this for an hour. I was vegetarian and didn't want to eat anything with vertebrae, especially if my grandmother had touched it. I suggested going to a restaurant instead, but my mother explained that my grandmother rarely left the house anymore. She was in her eighties and her legs were weak. Wouldn't it be a shame if she slipped and fell, I said, and my mother pinched me on the arm. She warned me to behave, and I explained that I always did. She rolled her eyes.

We sat on the floor, hunched over a low table. My grandmother was our only relative who still ate this way—the rest of my family switched to kitchen tables with chairs. I looked at the "feast" my grandmother had prepared: a large bowl of rice; gray, overcooked bean sprouts; kim chee that I knew would be bland because it lacked the bright red spices; a bowl full of something chunky and greasy—turnips, maybe. A whole fried fish, whose gummy, cloudy eyes stared at me despondently. I felt another pair of eyes on me, too. My grandmother was scrutinizing my face. The two dark pebbles underneath her drooping, wrinkled eyelids moved over my forehead and my cheeks. I ducked her stare and looked at the fish again. How could I put that in my mouth? I had been vegetarian for nine years; I had forgotten what fish even tasted like.

"Her skin is too dry. What is wrong with her? She should wear more make-up." My grandmother liked to talk about people in the room as if they weren't there.

"I guess it's the Seoul weather. It's very windy and cold here." My mother reached into her purse and handed me a jar of cream.

"Yes, it's the weather," I echoed and smiled apologetically at my grandmother. I dabbed the lotion on my face, which I thought felt a little greasy.

My grandmother's eyes moved over to my mother's face. "You're wearing too much make-up. What are you trying to hide? Wrinkles? You can't hide them forever."

"I'd like to try." My mother forced a laugh. I did too.

"Why is she so skinny?" My grandmother gripped her hands around my wrist. Her hands were surprisingly soft and warm. Apparently, the devil moisturizes.

"She's not so skinny. She's fine."

"I don't think you feed her enough. It's your cooking."

"She has a great appetite." My mother looked at me, glanced at my bowl, and looked at me again. I shoveled a spoonful of rice and some dingy bean sprouts into my mouth. The only bean sprouts I've ever liked are my mother's. She is an amazing cook, and her dishes are the first to go during potlucks.

"It's very delicious. Thank you." I picked up my napkin and wiped my mouth. Normally in front of other relatives, I would say more. But in front of my grandmother, my remedial Korean wrapped around my tongue and choked me. Or maybe that was the food. I studied my bowl of rice. How many bites would it take for me to finish?

"What is she wearing? What is that?" My grandmother stared at me. I stared at my mother. My mother stared at my grandmother.

"Pardon me? What do you mean?" My mother spoke for the both of us. She had asked me to wear something conservative, and I had. I wore a maroon sweater and a pair of black pants. She approved and even wore a similar outfit.

"She shouldn't wear so much black. Where is she going, to a funeral?"

I poured my grandmother some tea, and without thinking, filled her cup to the top. This is considered impolite in Korea, and I knew it. I might as well have flung the tea in her face and smashed the cup against the wall.

"I'm so sorry. I wasn't paying attention." I bowed my head and looked blankly at the cup.

"This is very rude. Is this how they do it in America? Are all children this disrespectful?"

"Please, I'm sorry." I kept staring at her cup. Why was I such an idiot?

My mother laughed nervously. "It's OK Annie, it's just an accident. The pitcher is too heavy for you. Why don't you show Grandmother what you brought for her?"

My mother had purchased a blue cashmere cardigan for my grandmother before we left for our trip. She shopped for days in order to find the perfect gift, and she had even paid full price. Since my grandmother was always cold, my mother thought the sweater would be appropriate. I presented the gift to my grandmother, using two hands and bowing my head.

"What's this?" She tore open the box, looked at the sweater, and threw it aside. "Probably made in China."

My mother blinked. "No, it's made in Italy. It's cashmere; it's warm."

"It's not real cashmere."

"Please, Annie would never give you a sweater that wasn't cashmere."

I nodded my head in agreement. "If you don't like it, Grandmother, I can buy a different one for you."

"Why would you buy another sweater that I won't like? I like the sweaters that I have. I don't need a new one."

"I'm sorry, this was my fault. I told Annie that you would like it." My mother picked up the sweater and started folding it. I sat quietly and simmered. I hoped that when my grandmother finally arrived in hell, she would find indoor plumbing and new, unfamiliar furniture, including a very high kitchen table. I rubbed my lower back; sitting on the floor was giving me scoliosis.

"Annie's Korean is shameful. You didn't teach her."

I shoved a spoonful of rice and turnips into my mouth to keep it busy. The self-righteous American fireball in me wanted to come to my mother's defense; my patchy Korean was my own fault, and it wasn't *that* bad—my listening comprehension was far more advanced than my speaking skills. But the obedient Korean girl in me knew to stand back silently. There was no point in arguing; it would only make things worse. I had a vision of my grandmother spontaneously bursting into flames and me coming to her rescue by emptying the brass kettle on her. My grandmother wouldn't be hurt too badly, just covered in her own piss, her body smoldering. She'd be grateful that she was still alive and finally realize what a horrendous bitch she had been her entire life. She'd probably smell better, too.

"Annie's Korean has gotten a lot better. I'm sure by the end of this trip she'll be writing books in Korean."

I tried not to laugh too hard.

"What would she write about? What does she know?"

In 1984, my grandmother turned seventy, which is a milestone in Korea, like turning sweet sixteen or fifty in America. My father and his five siblings organized a traditional Korean feast and family reunion in Seoul in honor of my grandmother. At the time, I was a diminutive eight-year-old with oversized teeth and an unruly perm. My mother had convinced me to get one a few days before flying to Seoul—perms were the hottest trends in both America and Korea. Curls helped limp American hair look fuller and helped straight Korean hair look more American—assuming American hair looked like black cotton candy.

"No, Anne, you hair so cute! You look like Shirley Temple!"

"No I don't, I look like Tina Turner!"

"You so silly. Everybody like you hair. They tell me, 'Oh you must be good mommy because you daughter so pretty.' Even you grandma will like."

"No she won't. Grandma doesn't like anything."

"Anne! Why you say such rude?"

"Because it's true."

Though I feared her deathly smell and her slashing tongue, I still had to pay my respect to my grandmother when she turned seventy. After all, she was my grandmother. It was the right thing to do, but even my father seemed filled with dread—his jaw tightened and the veins in his neck bulged the entire week before we left for Seoul. He had helped plan his mother's celebration, playing the part of the dutiful son, but from the look on his face I knew he wanted the party to be over before it even began. When I got older, I figured out that ever since my father had moved to the States, his relationship with his mother had been strained, and when he sponsored three other siblings and their families to immigrate, my grandmother never forgave him.

Relatives and family friends—some I knew, most I did not—donned their best *hanboks*, Korean traditional clothing, and packed into a banquet hall. The women wore floor-length, puffy skirts and colorful short jackets with wide sleeves that taper at the wrists. The men wore baggy silk pants with vests and decorated long jackets. Unfamiliar, clammy hands pinched my cheeks, patted my bottom, and stroked my hair, only to get their watches and rings caught in the frizzy black mass that engulfed my head. I spent most of the evening at my mother's side, smiling and nodding at distant relatives, trying to understand what they were saying to me. Even my mother had problems understanding some of them—a few guests spoke in thick country accents. I tugged at my curls, trying to straighten them, and tugged at the bright red skirt of my hanbok. The skirt, made out of stiff fabric, was too big for me and a good two inches dragged along the floor. The sleeves of the mint green and gold jacket were also two inches too long. The hanbok was a hand-me-down from a cousin and I needed another three years to grow into it. No one had passed down a pair of traditional white slippers, so I wore my Reeboks instead, and my feet were the only comfortable part of my body. My coarse, heavy petticoat irritated my legs. I reached underneath my skirt and furiously scratched them.

"Anne, stop! Why you so itch?"

"Because it's itchy."

"Everybody else not itch. Everybody stand still. You make Grandma mad! Stop!"

My grandmother, whose light blue, elegantly embroidered hanbok did little to soften her scowl, took her place at the long table in front of the crowded banquet hall. As everybody sat down to eat dinner, my mother led my brother and me to an empty dining room.

"After dinner, all grandkid bow to Grandma. You bow like this."

Instead of bending at the waist, with the arms hanging straight at the sides—the bow we used to greet and thank people—my mother raised her arms above her head, slowly went down on her knees, and sat on the backs of her calves. She gently lowered her hands on the floor in front of her and bowed her head solemnly between them. Her forehead lightly touched the ground and she held this position silently for a few seconds. The voluptuous skirt of her pale pink hanbok spread around her magnificently. Then she rose effortlessly and clasped her hands lightly in front of her. She paused dramatically, her eyes gazing piously at the floor. She looked up at my brother and me, her painted lips forming a smile that framed her perfect, white teeth.

"Remember? This bow we use for special day."

My brother and I had performed this bow several times before—it was the same bow we used during ceremonies to commemorate our dead relatives and the same bow we performed in front of our elders on New Year's Day to wish them prosperity. But I had never executed the bow in a floor-length-plus-two-inches skirt with a cumbersome petticoat underneath.

"OK, we practice now. Mike, you try."

Surprisingly, my brother completed his bow perfectly. Despite the generous layer of fat that insulated his middle and his stubby legs, he bowed rather gracefully, in one fluid motion. The pants of his gray and maroon hanbok offered more maneuverability than the stiff skirt of mine.

"My only son look so good! I think Grandma like very much! OK, Anne, you try."

As I began kneeling, I tried to avoid stumbling over my skirt, but I lost my balance anyway and ended up slamming my rear down to the floor with my feet splayed in front of me. Hoping my mother wouldn't notice, I swiftly lurched my chest forward and bowed my head. When I stood up, my skirt twisted around me, wringing tightly around my hips. My mother was mortified. She stared blankly at me, her mouth hanging open.

"Oh my GOD. No, Anne, oh no, go slow. Like this."

She demonstrated the bow again, slowly, so I could take in her graceful, calculated movements.

"Mom, that's what I did."

"No, I not know what you did. You bow look . . . crazy. Try again."

I started to kneel again.

"Don't forget arm!"

I raised my arms above my head and as I lowered down to the floor, my knees wobbled and I fell over on my side. My skirt flipped up to reveal my pasty, skinny legs. I heard my brother snicker. I flashed him the stink-eye.

"Oh no, no. Why you always fall? Bow not hard. Everyone can bow. Even baby can bow. Very easy, you know? Maybe you think too hard."

My brother erupted in laughter.

"Shut up, Mike!" I clenched my hands into menacing fists of destruction, but they probably looked more like cotton balls.

"Anne, Mike, be nice!"

"He's the one not being nice. He's making fun of me."

My brother mimicked my bow—flopping onto the floor like a beached whale and convulsing on his side.

"MOM, TELL HIM TO STOP!"

"MIKE!"

My brother looked up innocently. "What? I'm just bowing."

My mother pointed her finger to a chair. He gleefully sat down and watched eagerly.

"Anne, practice more. Think like ballet."

"But we don't kneel in ballet."

"No, no I mean you move very slow and very smooth. You see?"

She demonstrated again and slowly talked me through the process—raise the arms, kneel slowly, touch the hands to the floor, bow forward, count to three, stand up. There were too many steps to remember for my eight-year-old brain.

"Mom, why can't I just bow the other way? I keep tripping. The skirt is too long."

"That not right. This special bow. You grandma turn seventy. Very important age in Korea, you understand? Everyone bow like this for you grandma."

Finally after ten minutes of bowing, I managed to perform a few bows without falling over. I kneeled slowly enough so that I could use my hands to move my skirt out of the way, a cheating tactic I hoped no one would notice. The skirt's bulk made sitting back on my calves difficult, so instead I raised my rear in the air as I bowed my head forward. Again, a strategy I hoped would go undetected. The important part was that I didn't fall. I had all the poise of an intoxicated rhino, but at least I didn't stumble.

"I'm hungry. I want to eat."

My mother looked nervous. Clearly, I needed to practice more. My bow was more of an insult than a sign of reverence. I could sense my mother's regret for not having me practice bowing in my hanbok earlier. She fidgeted with her rings.

Though I didn't realize it at the time, my mother was under a lot of scrutiny because my brother and I were the only children in the entire family who were born in America, the only children with American first names on their birth certificates. My mother wanted my brother and me to make good impressions on the relatives and family friends we were meeting for the first time. Above all, she wanted us to impress my grandmother. As far as I know, my father's mother is the only person who has ever made my mother feel insecure. My mother wanted to prove that she raised well-behaved, Korean children in America. When we visited Seoul, my mother dressed my brother and me in our finest clothes, had us bear gifts for my grandmother, and coached us on flattering Korean phrases we could say to her ("You look so young" and "I love your house. It's beautiful."). Despite our best efforts, my grandmother remained cold and unimpressed. No matter how much my mother talked up my brother and me, we still couldn't hide our faltering Korean. My grandmother didn't care that we were both straight-A students at the top of our classes; we hadn't read the great Korean scholars in the language we were meant to speak. We were not Koreans. We were Americans. Executing a perfect bow in front of my grandmother and all my relatives would prove that even though my brother and I weren't completely fluent in the language, we were fluent in the culture. Of course, at eight years old, I didn't understand all of this. I was hungry.

"I'm tired of practicing. I want to eat!"

"Anne, you have practice more. You bow not look so good."

"I'll practice after dinner."

"No, after dinner you have bow for Grandma."

"Don't worry. I can do it. If Mike can do it, I can do it."

I heard my brother scoff. My mother was not convinced. "No keep practice."

"But I'm hungry. Aren't we gonna eat with everyone? Shouldn't we be in there with Grandma?"

She sighed and led us back to our table. I sat with my parents and listened to my relatives give speeches. My father's speech was filled with big words and complicated sentence structures. I could make out the words "luck" and "grandchildren" and "happiness." During our week-long trip, my father had remained tense and silent. Though I never heard an argument between my parents, I could sense that their relationship was strained by my grandmother's presence. Their playful sarcastic banter was subdued and my father smoked more cigarettes than usual. I think my father wanted to go back home. We all did.

While I ate my dinner, my mother kept arranging large napkins on my lap so I wouldn't spill on my hanbok. Finally, an uncle announced that all the grandchildren would come up and bow to their grandmother. When my oldest cousin approached my grandmother's table and bowed before her in front of nearly a hundred relatives and friends and waiters, I froze. It hadn't occurred to me that I'd be bowing in front of everyone. My heart tried to jackhammer its way out of my chest. I felt nauseous and the knotted curls near my temples became matted from cold sweat. I turned to my mother with panic in my eyes.

"I have to bow in front of everybody?"

"Shh, yes."

"I don't want to do it. I can't."

"No, Anne, you have to. Watch how everybody bow. You told me not worry. You can do it, yes."

Immediately I regretted not practicing my bow more. For the next fifteen minutes, I watched eleven cousins perform textbook-perfect bows. The grandchildren in each family came up and bowed at the same time, side by side. My grandmother returned their bows with a grave nod—not exactly a sign of approval, but not a sign of disapproval either. I watched as my cousins Yoon-chong, Yoonmi, and Woo-jay approached my grandmother. Yoon-chong was one of the oldest cousins, and her bow looked experienced. Yoonmi was an accomplished ballet dancer, and she bowed deeply and elegantly like a swan, with a perfect curve in her back. Woo-jay was majestic in his hanbok and his confidence came through in his bow. I glanced at my brother. His hanbok was a little tight around his doughy figure. But he seemed calm. He sat watching my cousins, lost in thought or boredom. His fingers pulled mindlessly at a string on his vest.

In my mind, I raced through the movements: raise arms, kneel, bow forward, count to three, stand up, don't forget to go slowly, and watch out for the skirt! I looked down at my lap and realized I had twisted my skirt around my hands and wrinkled it. I tried to smooth it out and wondered if anyone would notice me and my bow and my wrinkled skirt. Surveying the banquet room, I discovered that every single pair of eyes was locked on my cousins. Yes, everyone would notice. I started sweating and scratching my arms, a nervous habit further intensified by the stiff sleeves of my hanbok.

Finally, my parents pushed Mike and me toward my grandmother. I took a deep breath and started the 200-mile journey toward the front of the room. My throat was parched and I felt a nest of curls stick to my damp neck. We walked slowly toward our grandmother, who watched our every move, taking a tally of what we were doing wrong. My skirt rustled around me and I picked it up so I wouldn't trip. I looked over at Mike; he seemed comfortable and poised and maybe a little nonchalant, the way eleven-year-old boys often look. We approached my grandmother's table and I looked fearfully into her sunken eyes. Liver spots were sprinkled all over her cheeks and I could see her flaky scalp through her thinning, gray hair. My heart was beating so loudly I was sure everyone in Seoul could hear it. Out of the corner of my eye, I saw my brother lower himself to the floor, and I began my bow too. I raised my arms and began to kneel slowly. My entire body shook nervously as I lowered my arms and knees. I tried to sneak my hand down to push my skirt out of the way, but suddenly, my sneaker got caught in the jungle of ruffles in my petticoat, and I pitched forward. My body twisted to the side and my shoulder came slamming down to the floor. My skirt was wrapped around my legs and in a panic, I tried to shake them loose. I felt like a fish out of water, squirming and flopping in the final throes of death. Gut-busting laughter shook the entire room and my face flooded with humiliation, making me feel even more conspicuous in my bright hanbok. From the floor, I glanced over at my brother and realized I only saw his feet. He had finished his bow and was already standing up and absorbing the mayhem I had caused in the banquet room. I scrambled to my feet, but stepped on my skirt again and stumbled forward.

My grandmother's eyes stabbed me right through the heart. She did not nod. She remained silent while everyone else in the room slapped their knees and wiped their eyes. As my eyes filled with tears, I turned around and walked quickly toward my parents. I didn't stop; I continued passed them, toward the bathroom. My mother got up and followed me.

"Anne, what happen?"

"I don't know. I fell."

With trembling hands, I started taking off my hanbok, grabbing the petticoat and the skirt and throwing them on the floor. I wiped my eyes and nose with the back of my hand.

"What you do now? Put on hanbok!"

She grabbed my clothes and tried to put them on me. I went limp in her arms. She gave me a hug and rubbed my back. She handed me a tissue, chuckled gently, and tried to untangle my curls.

"Everyone laughed, everyone hates me."

"Who hate? Why they hate? It OK, Anne. It accident. The skirt too big on you. I should sew up shorter."

"But everyone laughed at me."

"Don't embarrass, Anne. It OK. Everyone still love you."

"Not Grandma. She's mad. She doesn't like me."

"Oh Anne, remember? Grandma not like anything."

She winked and led me out to the banquet room and my relatives grinned as I walked by them. They pinched me and told me how cute I was and that one day, I'd be able to bow like my cousins. I sat down between my father and my brother. I could tell Mike was holding back a barrage of witty comments; his pudgy face was ready to explode. But he showed remarkable restraint. My guess is that my father had told him to be nice. I searched for disappointment in my father's face, but there was none. It was blank. He passed me a sticky rice cake.

"It OK, Annie. Eat dessert and then we go. I think everybody very tired."

My grandmother never said anything about the incident, but I'm sure she blamed my parents for raising such an ungraceful American girl.

I looked across at my grandmother, who ate her lunch quietly. Our conversation, as upbeat as it was, came to an awkward silence. My mother, shifting uncomfortably on the floor, offered information about our family in the States: My aunt opened up another laundromat. A cousin got a job as a costume designer. My father's lab work was going well. My grandmother didn't seem interested. Then my mother talked about me: I graduated from Berkeley. I wrote and edited textbooks. I was interested in photography. My grandmother yawned, her warm breath practically melting my face. I tried to think of things to say that wouldn't offend her.

"Your house is so . . . nice. Everyone in Korea lives in apartment buildings, so this is special." I smiled weakly and looked around the house, trying not to stare at the puckered linoleum and a piece of long yellow tape that hang from the ceiling to trap flies and mosquitoes.

"Anne, wouldn't it be nice to live in a house like this?" My mother reached over and put some more bean sprouts on my plate.

"Give her some fish."

"Oh no, I'm very full. I can't eat anymore. Please." I shook my head and put down my chopsticks.

"You're too skinny. I thought everyone in America was fat."

My mother laughed lightly.

"I said, give her some fish."

I looked at the fish in despair. It's side had been split open to reveal white and gray flesh and part of its spine. One of its gills had been yanked open to get to the meat underneath. It looked greasy. My stomach churned.

"I've already had some, and I'm very full." I rubbed my stomach and looked at my mother in a panic.

"Don't lie to me, you didn't eat any. I've been watching. Eat some fish."

"Oh, but there's only a little left. We've been eating it. It's very good. The fish in Korea is so fresh." My mother helped herself to more fish.

"Why isn't she eating the fish? Does she think my cooking is bad? Does she think her mother's is better? Is that why?"

"No, no . . . I'm so full . . . but I can eat more." I reached over and tore off a piece of fish and popped it in my mouth.

My grandmother's fish was saltier than salt. I chewed carefully, trying not to involve my tongue, whose taste buds were wilting and dying. The texture was squishy and slimy and I felt as though I was eating a leech. As a vegetarian, I was accustomed to fibrous plant matter, foods that only took a few chomps and a swallow. But the fish was like gum; no matter how long I chewed, the salty morsel in my mouth seemed to stay the same size. I wasn't sure if and when I should swallow. My mother looked at me, trying not to look too surprised.

"Look how much Annie likes your fish."

"It's so delicious." Chomp, chomp, chomp. I reached over for more, but my mother beat me. She tore off a gigantic piece and moved it to her rice bowl.

"You're taking all the fish, Mom. Leave me some." I picked at the carcass and found a small sliver of flesh.

"I can make more." My grandmother looked satisfied, finally.

"Oh no, you don't have to do that. We have plenty here." My mother shook her head.

"Please, no you've done enough, Grandmother. I can't possibly eat anymore. Now I'll be fat like an American." I couldn't tell if she smiled or not. I wasn't even sure she could smile.

When my mother and I finally left my grandmother's house, we searched the block frantically for a store—we both needed to use the restroom. We found a convenience store and my mother bought me a grape Fanta to justify using the facilities. When I visited Seoul as a kid, grape Fanta was my favorite; it was a soda not widely available in America. But as I took a sip, I realized it was much too sweet for me. Still, I took big gulps to get rid of the fishy, sour taste of respect in my mouth.

FOOL WHO PLAY COOL

"Anne, wake up."

I felt a poke in my side but refused to acknowledge it. There was no light coming through the window. Therefore, it was not an appropriate time to wake up.

"Anne, I say, wake up."

I whimpered and turned onto my side, which was particularly painful because I was sleeping on the floor. I've heard that doctors recommend sleeping on the floor because it's good for the back, but I have yet to hear about a doctor who actually does it. In the middle of the night, half my vertebrae had fallen out. My mother shook me vigorously.

"Go away!"

"Wake up! I pinch if you not wake up." She tugged on my hair and stuck her finger in my ear, a tactic I used on her when I was little and wanted her to wake up and take me to Disneyland. I also used to pry her eyelids open, pinch her nose so she couldn't breathe, and begin dressing her while she was still in bed. I swatted her hand away.

"Please, woman, go away. What time is it?"

"You uncle wait for you! Wake up!"

"What time is it?"

"Don't worry about time. Time for wake up." She threw the covers off of me and jerked away my pillow.

"MOM!"

"Shhh, you wake up you aunt."

"Why does she get to sleep? What time is it?"

"Hurry up and get clothes."

"Why?" Without my glasses, I squinted to see the clock. It was 4:15. In the morning. "What the hell? It's four!"

"No, it five!"

"No, it's not, it's four. Four-fifteen."

"See? Almost five, time to wake up!"

"No it's closer to four and since when is five a time to wake up?" I grabbed the covers and put them over me again. I curled into a tight ball and wrapped my arms around my head, the position I learned to take during earthquake drills in elementary school. It was only my third day of a two-week trip to Korea with my mother and I was already plotting to push her into the Han River.

"Why you so grump?"

"Because it's *four o'clock in the morning.*"

"We have to go!" My mother threw off the covers again and then yanked up a corner of the blanket I was sleeping on. I rolled off.

"*Moooommmm!*" I grumbled and stood up. "Where are we going?

"We go Soraksan. We leave now."

I groaned. Soraksan is a mountain and a national park. I remember going there when I was five and crying because my legs were tired. Even then I knew that hiking is the devil's work. "You didn't tell me we were going there."

"I tell you we go."

"No you didn't."

"Yes I do. Remember yesterday I say, 'Anne, tomorrow we wake up early go Soraksan.'"

I thought about it for a second. No, she most definitely didn't tell me. I think she has a lot of conversations with me in her head, in a fantasy world where she talks and I listen and nod my head silently. I would never agree to wake up early to do anything except for sleep. "You never told me that. Where is it?"

"You don't remember?" She threw a pair of pants at me. "Anne if you not get ready now Mommy get very annoy."

"You're already annoy." I started to put on my pants but stopped to rub my ass; my left cheek was asleep, just like the rest of me should've been. My mother had told me that eventually I'd get used to sleeping on the floor and develop a Korean's backside, with the ability to sleep on the hardest of floors, even on a slab of frozen granite. My ass ached; I guess it was still American. I yawned and felt a shirt hit my face. "Look, I'm getting ready. Where is Soraksan?"

"Near Sokcho." She threw a pair of socks at me.

"These are dirty." I threw them aside. "Where is Sokcho?"

"Why you not know Soraksan and Sokcho?" She threw another pair of socks.

"Because I'm not from here. Where is it? These are your socks."

"North and east, near ocean. We go mountain and you see temple for Buddha. Wear my sock."

"Why do I have to go see a temple? I don't want to wear your socks, that's gross."

"You see tree too. Many, many tree. Why gross? They clean! We have to leave."

"Because you might have fungus or something."

"Anne!"

"OK, OK I'll put them on. Why do we have to leave now?"

"Because it far away. Five hour."

"Are you kidding me? Just to see a temple and some trees? There are closer temples—they're all the same—and there are trees outside. I can see them from here."

She pushed me into the bathroom and I stepped in a puddle. I groaned. My mother's fungus-socks became completely soaked. Korean bathrooms tend to get very wet. There's usually no curtain or glass door that separates the shower/bathtub from the toilet and sink—everything just sits together in a tiled room with a drain in the middle so one can make a watery mess or hose down the entire floor. I always find Korean bathrooms a little unnerving. I'm used to bathing in a smaller space and I worry that I'll get my towel wet. I peeled off the wet socks and hung them on the towel bar, which was also wet.

My mother knocked on the bathroom door. "Anne! Everyone wait!"

I wrenched the toothbrush out of my mouth and met my mother in the kitchen.

"What happen to you sock?"

"I can't wear them. They're wet."

"Anne I tell you, no fungus." She scowled and handed me a piece of heavily buttered toast.

"This is breakfast? What about Korean food?" For the most part, people in Korea eat the same thing for breakfast, lunch, and dinner. My stomach and I find this very pleasing. "I'm in Korea; I want Korean food."

"It Korean toast."

"Are you trying to be funny?"

"We get Korean food later. No time, everyone wait."

"Who's everyone?" I looked around the kitchen. "There's no one here."

On cue, my uncle walked in with a gigantic smile; he is a morning person, just like my mother. "Good morning, Annie." My mother's brother-in-law is a short, pear-shaped bald man with moles in awkward places and crooked, dingy teeth, but he has a boyish, cherub-like quality to his face. He makes a lot of silly jokes and always wards off relatives whenever they give me a hard time. I gave him the customary bow and morning greeting. I lightened up; going on a five-hour trip with my uncle could actually be fun.

"Did you sleep well?" he asked me in Korean.

"I'd like to sleep more," I answered in Korean.

My uncle laughed. "Aren't you excited for the trip?"

"Sure. Are you?"

"No, no, I'm not going."

"You're not? But you have to come!"

"Your aunt and I have to work, sorry, my little Annie. I'm dropping you off at your mother's friends' apartment. They're taking you to Soraksan."

"Which friends?" I like some, not all, of my mother's friends. I looked at my mother suspiciously.

"Eat, Anne." She pushed the toast toward me. "Hurry up and we go." She dug around in her gigantic leather purse and pulled out a pair of socks.

"You carry socks with you? I have socks in my suitcase, I can get socks." My mother ignored me and stooped down to jam the socks on my feet. "Are they clean at least?"

"Anne . . ."

"If you had woken up earlier you could've eaten Korean food with me," my uncle nodded apologetically. My mother looked up at me, smug. I rolled my eyes and ate my toast. It wasn't buttered; it was margarined. My family, like the rest of Korea, loves margarine and always refers to it as butter, even though I've explained that it's not.

December in Korea is cold and windy and as we walked outside my aunt and uncle's apartment building, the late night/early-morning gale whipped through my jeans and froze my underwear to my sore ass. I shivered.

"Where you scarf?"

"I left it inside."

My mother scoffed and tried to give me her scarf.

"I don't want your scarf." I grit my teeth against the cold. "I'll be OK."

"Don't be stupy." She dug in her bag and brought out another scarf.

"What do you have in there?"

"You want hat? I have hat."

"No, I'm fine."

She wrapped her scarf around my head and laughed. Her scarf was warm and smelled like her perfume, Eternity by Calvin Klein.

"You look like Grandma."

"Thanks."

In Seoul, most people live in clusters of high-rise apartment buildings. The residents park their cars in tight rows between the buildings and leave their cars in neutral. An attendant pushes the cars around so that people can drive out or park. When I was young I thought that being a parking attendant was tough work, but then I figured out that everyone drove Hyundais, which have bodies made of soda cans and engines fueled by hamsters that run around in wheels. With a burning cigarette dangling from his mouth, the attendant pushed away the cars around my uncle's champagne Hyundai. My mother took shotgun.

"We go with Gi-sook and her husband. You remember Gi-sook?"

"No."

My mother paused to think. "Hmm, maybe you too young. Mommy friend. Old, old friend, from before you born. She and her husband is farmer."

"Farmers? What do they grow? Rice?"

"No, no, pharm-ah-, you know they give out medicine."

"Oh, you mean pharmacist."

My uncle nodded and repeated after me, "Fah-ma-seet-uh."

"Close enough."

"When was the last time you went to Soraksan?" my uncle asked me in Korean.

I shrugged. My mother answered for me. "She must've been six or seven years old. So about fifteen years!"

"I don't remember much."

"You go see Buddha over summer, remember?"

"All I remember is taking piano lessons."

I spent most of the summers in Seoul taking piano lessons. My mother hoped that I would become a classical pianist like my cousin even though I displayed no real talent or interest. She thought that if there was anyone who could squeeze Beethoven out of my tiny stiff fingers, it was a Korean piano teacher—my American teacher certainly wasn't doing an adequate job. My Korean teacher swatted at my hands and gave me daunting sheet music with black dots splattered all over the page. She even sent me to an herbalist to get a special tea designed to strengthen my hands and fingers. I remember the tea smelled exactly like garbage so it was very counterintuitive to swallow. Still, I drank it everyday and my fingers remained limp and lifeless. Ever since then I've been wary of Eastern medicine.

"Such waste, you piano." My mother shook her head.

"I wasn't *that* bad. I just wasn't good."

"Same thing, Anne."

We arrived at a pharmacy, at the bottom of an apartment highrise. A big green plus sign hung in the storefront. A black Hyundai waiting outside honked twice. My mother and I dashed out into the cold and scrambled into the backseat. A man and a woman, both around my mother's age, sat in front. My mother grabbed each of their shoulders and squealed happily. It was the first time I had ever heard my mother squeal. I've heard her grunt, scoff, growl, groan, moan, yelp, and even snarl, but I have never heard her squeal. I winced. It was much too early to squeal.

"It's been too long! How have you been?" My mother poked her head between the front seats so she could get a better look at her friends. "You're so old! What happened?" The last time my mother saw her friends was when she visited Korea a year ago.

The woman smacked my mother playfully on the arm. "Hello! Hello!·Oh·you·must·be·so·cold·it's·so·cold·outside· I'm·pretty·sure·it·doesn't·get·colder·than·this·the·numbers· don't·go·low·enough·how·could·this·little·country· of·ours·get·so·cold·you·know·every·year·I·forget· how·cold·it·gets·and·then·winter·comes·and·I'm·always· so·surprised·really·it's·so·silly. . . . " She looked at me, smiled, and took a deep breath. "Wow·this·is·Annie?·I·haven't·seen·

you·in·oh·I·don't·know·how·long·too·long·you·don't· remember·me·do·you? You·were·this·high·up·to·my·knee·or· maybe·even·smaller·though·you·look·pretty·small·now."

I noticed that the ajuma liked to repeat things twice, which improved my comprehension, but her words whirled around my head like a swarm of locusts that devour everything in its path. I put on my seatbelt and bowed awkwardly and desperately tried not to stare at her face. Something was a little off, a little unsettling.

She had no eyebrows.

Her eyes, nose, and lips all seemed very far apart from each other, each floating in a vast landscape of pale, flawless skin. The only things that separated her eyes from her wide forehead were two thin, curved lines she had drawn with a brown pencil. She looked perpetually surprised. I smiled uncomfortably.

The woman's husband turned his head and looked at me. "Annie doesn't remember us, does she?" Koreans like to address children in the third person right to their faces.

I shook my head. "No, I don't, *Ajeoshi*." Koreans address anyone older as auntie (ajuma) or uncle (ajeoshi), even if they aren't related. It's nice because you never have to remember names.

"When you were little all you did was cry. It was the cutest thing. You were made of tears and snot." The Korean word for *snot* translates literally to "nose water," which sounds a lot nicer than snot. "You cried all day and all night, forever and ever, for all eternity." The ajeoshi looked at my mother. "Does she still like pickles? I remember that was all she ate. Salty, spicy, pickled food."

"I still like pickled food, and I finally stopped crying last year," I said in Korean. The adults laughed. My mother winked at me. She liked when I charmed her friends.

While the car slowly made its way through morning rush hour traffic, the adults in the car chatted noisily in Korean and I struggled to keep up. Once I get tripped up on one word, I miss everything that comes after it and then I hopelessly try to piece bits and pieces of the conversation together again: Something about a pig or an actor, or maybe a pig-faced actor, who opened up a store or restaurant, maybe a bank or zoo, with bad service or maybe someone went on strike, and then something about oil or was that gasoline? They talked so fast that it was impossible to figure out when they moved on to a different topic. Occasionally I'd hear my name pop up: Korean, Korean, Korean, "Annie," Korean, Korean.

"What?"

"Nothing." Chuckle.

"What'd you say about me?"

"Nothing." Laughter. Korean, Korean, "Annie," Korean.

"What? Tell me."

"No, we not say anything."

"Liar," I said in Korean. Everyone laughed. "Don't listen to her; she's full of lies."

"That's no way to talk to your mother," my mother answered me in Korean.

"Yes, listen to your mother," echoed the ajeoshi, "Everyone should listen to your mother, even if she's a liar." The ladies giggled.

"Oh-she's-so-cute-and-how-funny-and-how-adorable-just-adorable-and-what-a-beautiful-coat." She reached over to pet my coat.

"You like it? It's Stefanel. I picked it out," My mother smiled proudly. On our first day in Seoul, my mother dragged me shopping to buy a proper coat, "the kind that real people wear." It was long, grey, and had one button and a sash. It looked like a bathrobe. My mother explained that it wasn't a bathrobe; it was Italian. "Annie doesn't like it."

"I didn't say I didn't like it," I started to protest but decided to just leave it. We still had four and a half more hours to go.

"It's·beautiful·I·guess·this·is·what·the·young·kids·are· wearing·these·days·you·know·my·daughter·goes·shopping·all· the·time·I·think·she·lives·at·the·mall·and·she·has·tons· of·clothes·mountains·of·clothes·but·she·always·wears·the· same·thing·I·think·she·buys·the·same·thing·over·and·over."

My mother laughed. "My daughter wears the same thing as homeless people, except she pays for them, and she wears her pants until there are holes in the crotch and the rear. How do you get holes there? It's a mystery."

"MOM!"

Whenever I spend time with my mother and her friends, she always recounts every single thing I ever did wrong and discusses every embarrassing detail about me and then her friends describe all of their children's shortcomings and the whole conversation degrades into a contest to see who has the world's most lazy, disrespectful, loud, or defiant kid. Normally Korean mothers talk up their children, but when they don't have to impress anyone, they swing the other way.

"My·daughter·talks·on·the·phone·every·night·for·hours·and· hours·and·she·talks·to·friends·she·just·saw·all·day·what·could· she·possibly·have·to·talk·about·nothing·has·happened·since·she· saw·them·last·the·phone·is·like·a·drug·she·can't·stop."

"Well, you're a pharmacist, so you should be able to find a cure. But at least your daughter is *yamjonheh*, my Annie could be more yamjonheh."

In Korean, the word *yamjonheh* describes someone who is obedient and modest—someone who is courteous and docile, and maybe a little bookish and shy. Girls who are yamjonheh cover their mouths when they laugh. Korean parents would like their kids to be yamjonheh. This doesn't always work out. When I laugh, my mouth becomes a gaping hole in my face and rice comes flying out.

"Kids are horrible." The ajeoshi shook his head. "Too bad you can't give birth to adults." He laughed and slapped the steering wheel.

"We were never this bad when we were young," my mother declared. I rolled my eyes. I've heard her mother say the same thing.

As we sped along the highway, the concrete and glass of Seoul melted away into a rural landscape. Small homes popped up in the middle of large muddy fields and I could see fog-topped mountains in the distance. Seoul is like Los Angeles; it's easy to forget that there are mountains outside of the city. When my mother was in high school and college, she liked to camp, hike, and rock climb. It's hard for me to imagine my mother rough it in the wilderness, sleeping next to the insects, snakes, tigers, and whatever else lives in the Korean bush. I guess she's more rugged than she seems. I happen to hate camping; nature is best viewed from indoors in a safe, allergen-free, and climate-controlled environment with wireless Internet.

Jet lag plus an early start to a morning is a troublesome combination. Though I tried to keep up with the conversation in the car, I eventually drifted off into my own thoughts and then drifted off to sleep.

I woke up with my mother's finger poking my ribs.

"Wake up."

"Huh?" Déjà vu. I looked at my watch. Only one hour had passed.

"Break time. Food."

We were at a roadside food court that struck me as very American. There were plastic tables and chairs in the middle of a large cafeteria that was lined with different fast-food vendors, who urged customers to their counters. I settled on watery noodles and sat down. My mother got small greasy pancakes, sticky rice cakes, a bag of kettle corn, roasted chestnuts, dried cuttlefish, and a baked yam. Normally she eats healthier, especially after she got cancer, and nearly everything that enters her belly is low in saturated fat and made from whole grains, but she was on vacation.

"You're eating that? You can't eat that. Your food has no theme. It's all over the place."

"No I eat." She offered me a pancake. I took it gingerly with my thumb and index finger and kept it a foot away from me, careful not to stain my new Italian bathrobe with grease. I took a bite. It was filled with sugary red bean. I could actually taste the crystals of refined sugar melt my molars.

"Oh man, this is too sweet. It tastes like a donut. I can't eat this."

Every Saturday at Korean school, the students were served Winchell's donuts and Sunny Delight, a healthy start to a day filled with grammar, vocabulary, and boredom. The mere thought of deep-fried rings of fat or neon-orange citrus punch makes me gag.

She offered me a bite of her yam. "You know, when I was young, I eat this all the time. Mommy favorite. Everyday after school I walk home and I see ajeoshi who sell yam in little cart. I eat two, maybe three." She took a gigantic bite of her yam.

"They have yams in America, too."

"Mmmf mmmf mmmf."

"What?"

My mother swallowed. "No, Korean yam is special. Better." She looked at her tray eagerly, uncertain which to eat next. "Mommy so happy! I eat this all when I grow up in Seoul. I eat and eat and eat and never get fat. Not like now." She patted the spare tire around her middle.

"Oh·you·got·chestnuts·I·got·chestnuts·too·do·you·eat·chest-nuts·in·America·I·hear·Americans·don't·like·red·bean·Annie·do-you-like-red-bean?" The ajuma and ajeoshi sat down at our table with their trays.

"It's OK. Too sweet for me." I looked down at my noodles. They seemed so boring compared to what everyone else was eating. The ajeoshi was eating a very adorable lunch set with a tiny bowl of steaming rice and tiny dishes filled with tiny pickled vegetables. There was a tiny napkin and a tiny cup of tea too. My mother, in the middle of her snack attack, took little bites of everything in front of her.

"Noodles?" I offered them to my mother. She shook her head, her mouth full of cuttlefish.

My mother stuffed the rest of her snacks into her purse and we scampered back into the car and hit the road again. Then, to my horror, the adults started the singing portion of the roadtrip: Korean traditional music with yodeling vocalists, Korean pop stars, Korean oldies, and then more traditional vocalists. My mother sang along in her exaggerated church choir vibrato. Whenever my mother sings, she sings with gusto, even if she doesn't know the words. Sometimes she makes up words or mumbles and then throws in a few extra *oh*'s and *ah*'s at the chorus. My mother doesn't have a bad voice, but she sings with the confidence of someone who has a great one. My mother sang along and bobbed her head to a rousing boy-band number.

"How do you know these songs?"

"Korean radio. MTV Korea."

"You watch MTV? Dude, who are you?"

The adults continued chatting and singing and occasionally asked me questions. What did I study in school? What did I do for a living? Did I want to live in Korea?

"I couldn't find a job here," I answered in Korean.

"You could teach English." The ajeoshi fumbled with the CD player.

"Yes, you should teach English," my mother agreed.

"A·lot·of·Americans·teach·English·here·I·think·they·make· good·money·you·can·live·here·and·teach·English·how·fun· wouldn't·that·be·fun?"

"I don't want to teach English."

"Why·not? You·speak·English·you·speak·better·English·than· Koreans."

"But I don't want to teach it."

"You so silly, Anne, why you not want teach?" my mother asked in English.

"Then why don't you teach Korean in America? You speak Korean," I answered in Korean.

"Oh, she got you there. Your daughter's very smart. She must take after you." The ajeoshi smiled at my mother.

"She·can·teach·English·here·and·then·you·can·live·with·her· fun·right?"

My mother and I both winced at the thought.

"Break time. Let's get some snacks." The ajeoshi pulled the car off to another roadside food court.

"Again? We just had a snack," I answered in Korean. "I'm full. I can't possibly eat anymore."

"We'll·just·have·a·small·snack·maybe·some·ice·cream·do·you· want·some·ice·cream?"

"It's freezing!"

My mother and her friends got soft·serve and coffee. Everyone in Korea seems to love soft·serve, which is pretty much the margarine of the ice cream world. When we were young, we used to drive through McDonald's and my father would holler into the metal speakerbox and order three soft·serve cones and a milkshake for my brother. My mother lapped up her cone as she walked around another food court with me.

"What you want to eat, Anne? Mommy can buy you ice cream."

"It'll make me sick. It'll make you sick, too."

"But Mommy don't care. How about Polapo?"

I perked up. Polapo was my favorite as a kid in Korea. It was more or less frozen grape juice in a long paper cone that you squeeze into your mouth. "Do they still make it?" It stains viciously and my mother used to hate buying it for me.

We opened freezers and looked through the treats. She handed me something frozen. "I can't find Polapo, but you like this remember?" I looked at the label; it was a red bean popsicle.

"No, *you* liked this. You always got this and I always got Polapo."

"Really?"

"Yeah, and Mike got the chocolate ice cream that comes in a big cup with the wooden spoon on the side, and he'd use that to flick stuff at me. And Dad would get coffee ice cream."

"Then what you get now?"

"It's too early for ice cream. It's like eight o'clock."

"Why you no fun?"

We drove onto the highway again and an hour later we took yet another break, this time for gas and more cuttlefish and soda.

"We're stopping every hour. We're never going to get there! How much longer do we have to go?"

"On roadtrips in Korea, people stop every hour," the ajeoshi informed me.

"Really? But then it takes a long time to get anywhere."

"That's the point."

"That's so Buddhist," I whispered to my mother in English.

"Good because we go Buddhist temple."

"It's such a small country this little country of ours maybe we stop every hour to make it seem bigger but you know not as big as America."

In Korean school I learned that South Korea is approximately the size of Indiana. When I moved to the east coast, I drove through Indiana in a day, at ninety-five miles per hour, stopping only for Taco Bell.

An hour later, our car pulled into a small, bustling town on the coast.

"Are we here?" I looked out the window. People were milling around carrying large parcels and Styrofoam ice chests. Vendors were sitting on low chairs with their wares spread in front of them on a blanket. We were in the middle of a flea market. "This doesn't look like a mountain."

"This is Sokcho."

"How far to Soraksan?"

"Why you worry about how far? We in Sokcho now. You think about Sokcho."

"It's-time-for-lunch-I'm-so-hungry! What-should-we-eat-let's-walk-over-to-the-fish-market-doesn't-that-sound-like-fun-to-walk-over-to-the-fish-market-and-eat-some-fish-you-know-Sokcho-is-famous-for-its-fish."

"It's lunch time?" I looked at my watch, almost noon. "I can't possibly eat anymore."

"You find room, Anne. You know what everyone say, 'There always room for food.'"

"No, you mean, 'There's always room for dessert.'"

"You want dessert?"

"No, no I don't want dessert. I don't want anything. I can't eat anymore. I never want to eat again."

"You grump. Like crab."

"No, I'm full. Like a refrigerator."

We walked along the beach, watching the waves crash and shivering in the cold. We stopped at a street vendor to get some tea, which tasted fishy. In fact, everything in Sokcho seemed fishy. Stores and street vendors sold every form of fish imaginable: fish fillets, fish stew, fish sticks, fish balls, fish cakes, and shredded dried fish snacks. The heavy scent of fried fish wafted out of every restaurant, which showcased their fish in large aquariums, and gulls picked at random fish guts strewn in the streets. We walked to a lively fish market where fishmongers plucked sea life out of their nets and into large plastic tubs of water on the floor. The varieties were endless: small, big, short, long, flat, chubby, scrawny, or stubby with fins, feelers, tentacles, shells, or spikes in brown, pink, spotted, striped, and iridescent colors. All were alive, writhing in their bright orange tubs, waiting for their doom. Customers peered into the tubs, pointing to the exact specimen they wanted and asking the best way to prepare it at home. The answer was always the same: fry it.

I watched as my mother examined different tubs and discussed its contents with her friends. Whatever they pointed to, an old woman in a fish-smeared rubber apron scooped it up and put it in a separate tub with no water. Fish and shrimp flopped around, gasping for water, and mussels and clams slid around the bottom of the tub. Squid entwined their tentacles together, hugging for dear life in the final minutes before their dry, parched death.

My mother and the ajuma bickered over who would pay for the fish. They elbowed each other away from the fishmonger with their wallets in hand. My mother was victorious, but I saw the ajuma sneak in a few won into my mother's bag.

"Uh oh, Anne, nothing vegetarian for you." My mother shook her head in disappointment, not because there was nothing vegetarian, but because I was vegetarian.

"It's OK, I'm not hungry." I looked at a tub full of squirming black eels. They looked like hair.

"Oh-you're-vegetarian-but-you-can-eat-fish-though-right?"

I sighed. I knew where this conversation was headed. "No, no fish."

"How can you be vegetarian?" The ajeoshi stared at me incredulously. "What can you even eat?"

I forced a patient smile. "There's plenty of things to eat, just not here, but that's OK. I'm really not hungry." I remembered my grandmother's salty, rubbery fish from the day before and rubbed my stomach.

In the fish market, we sat down on milk crates around a small barbecue, which was more like a large metal soup pot with a grate on top than an actual barbecue. The ajeoshi lit the coals and wood on fire and our waitress threw shrimp on the barbie. Live. I watched as legs and antennae danced on top of the flame. The shrimp made faint hissing noises as steam escaped from their writhing, burning bodies.

"They're suffering." I watched them, horrified.

"No, they not suffer, they dance, see?" My mother laughed and pointed to a shrimp that was twisting and shouting on top of the hot grate.

"They should've cut off their heads first."

"Then it not taste as good." My mother used her chopsticks to flip over a shrimp. Its grey shell had turned bright pink.

The waitress dumped squid, mussels, and clams onto the grate. Squid tentacles curled up like ribbons and their translucent bodies turned white as they cooked to death. Luckily the shellfish didn't suffer and die as cinematically.

"You sure you not want try?" My mother plucked a squid off the barbecue with her chopsticks and dangled it in front of me. "It taste so good!"

I waved my hands and cringed. "It looks chewy. I don't like chewy things."

"You like gum. When you little you eat so much gum, you get cavity. We go dentist every week. I worry you teeth fall out."

"Gum isn't made from an animal."

"How come you not want to hurt animal but you hurt Mommy? You always make Mommy heart very sick!" She clutched her heart and pretended to cry, using her napkin to wipe away imaginary tears. Then, to make sure she had an audience, she translated her joke into Korean. Her friends laughed and I shook my head. She has been making that same joke ever since I turned vegetarian, and it was funny only once.

The waitress brought out the next course—a ceramic pot of bubbling fish stew. Fish heads and tails and shells bobbed around in a brown broth with mushrooms and onions. She put the pot right on top of the barbecue and started serving.

"No need to give one to my daughter. She's vegetarian. She loves animals too much. More than her own mother."

"It's OK, there's no meat in here." The waitress served me a bowl of fish stew.

After lunch we walked around the beach a little more. I shooed away the hopeful gulls that followed us, in search of a snack.

"How·many·times·do·you·think·we·came·here·together· when·we·were·young?   You·know·Annie·your·mother·and·I· used·to·come·here·all·the·time·and·I·don't·remember·what·we· did·there's·not·much·to·do·here·but·eat·fish·but·we·always· had·such·a·good·time."

"I think we came up at least ten times—came up here for the weekends and went to the beach. Remember how skinny we were?" She and the ajuma linked arms and smiled. Girls in Korea always link arms when they walk together and they lean their heads in together to whisper secrets and giggle, creating an intimate republic of two.

"Don't-you-miss-Korea? Korea-misses-you."

"I miss it all the time. When I first moved to America, I cried all the time because I missed everyone, everything. I missed crackers. Did you know that my mother mailed me a box of food every week with directions on how to prepare each thing? There were no Korean markets there back then."

"But-now-that-your-kids-are-old-they-can-take-care-of-themselves-they-don't-need-you-so-you-can-move-back-how-fun-would-it-be-you-can-live near-us!"

"No, no, I couldn't. Seoul isn't mine anymore."

We got back into the car and left Sokcho and all of its entrails behind. Less than an hour later, the ajeoshi parked the car again.

"We're stopping again? What is there even left to eat?"

"Anne, we here, silly."

I got out of the car and stretched and yawned. Just as my mother had promised, there were trees. Pine trees. I immediately thought of three things: a Korean restaurant in the San Fernando Valley called Pine Tree; *shikhye,* a sweet Korean drink made with rice and pine nuts; and pesto. The thought of more food made me nauseous.

"First-we-should-take-Annie-to-see-Sinheungsa-she-has-to-see-that-it-would-be-a-crime-to-come-out-all-the-way-here-and-not-see-that-a-real-crime-we-might-get-arrested."

"What is Sim-noon-gah?"

The adults laughed.

"No, Anne, Shhiinnn-hoong-saaa."

"That's what I said."

"*Ayoo,* don't lie to you mommy. Sinheungsa is temple."

We followed signs along a dirt and pine needle–covered path to the temple and passed by a three-story-high bronze statue of the Buddha, sitting cross-legged. His eyes were closed and he had a jewel in the center of his forehead. He also had really long earlobes and was wearing a one-shoulder toga.

"He must be cold." I smirked.

"Anne, it Buddha, you show respect." My mother rolled her eyes and tried not to laugh. "But I think maybe he cold."

"It's the largest seated bronze Buddha in Korea," The ajeoshi informed us, "I remember when it was built."

I looked at the statue. I had assumed it was thousands of years old, only because I always assume anything large, bronze, and religious is thousands of years old. "When was it built?"

"I think construction started in 1987."

"Really? That was practically last year."

Korea, like other Asian countries, has a long history of Buddhism—so long that I think many Koreans, especially the younger generations, see Buddhism as a part of the culture and history of the country and less as an actual religion that they practice. Even though my family is Catholic, we still honor our dead grandfathers with a Buddhist ceremony. We cook a special feast that includes sticky rice cakes and fruit, and then someone looks up a Buddhist prayer in a book and writes it on a piece of a paper. We bow several times and then turn away from the table, so that our grandfather can "eat" in privacy. Later we burn the prayer and dump the ashes into the gutter. When our family became more active in the Catholic church, we added a bible reading and sang a hymn after the Buddhist ceremony. My grandmother actually practices Buddhism and is a leader at her temple, but she didn't seem to mind when our family became Catholic. "Religion is religion," she explained, "Do whatever makes you happy." It was actually very Buddhist of her.

We walked through a gate with four ten-foot-tall, brightly painted statues of men, though they looked more like monsters than humans. One had a dark gray face with full cheeks, wide-set red eyes, and a spiky, white beard that framed his thin red lips. The other had a bright red face with a spiky black beard and large bushy eyebrows. He was wielding a long sword. Their flared nostrils and the deep creases in their faces made the men look as if they had smelled something unsavory and were angry for it. "Who are these guys?"

"I don't know." My mother translated my question into Korean for the ajeoshi. He shrugged. "They're statues."

"Why are they mad?" I asked in Korean.

"Maybe-they're-mad-because-they-don't-want-people-to-come-here-you-know-scare-away-people-though-I-guess-it's-a-temple-so-you'd-think-they-wouldn't-be-angry-but-more-like-hey-come-in-have-some-coffee-though-I-guess-monks-drink-tea."

Despite the rather cold welcome, we walked through the gates and into Sinheungsa. I was surprised; it looked more like a compound than a Buddhist temple. There were several low buildings with tile roofs and brightly colored moldings of dragons and geometrical patterns. There were stone pagodas that stood between the buildings, and several stone staircases that lead to small shrines with golden statues of the Buddha and smoking pine incense in round brass cups filled with sand. There were also living quarters and meeting rooms for monks, and their white shoes were lined up outside the heavy wooden doors. When I turned vegetarian, my mother joked that she would send me to live with the monks and eat their food, which contains no animals or flavor.

A chilly breeze blew right through us and my mother rummaged in her purse and offered me a fuzzy black hat and matching gloves.

"What about you? Do you have some for yourself?"

"No, Mommy not cold."

"Liar. Let's share—which one do you want?"

"No you take both. It set."

"That's retarded." I jammed the hat on her head and shoved my fingers in her gloves. They were too big. My mother has long, thin, elegant hands. My thumbs are freakishly small so I usually wear children's gloves.

"OK·are·you·ready·to·go·up·the·mountain·Annie?·The·view·is·really·beautiful·it's·really·something."

I forced a smile. "You know, if you're tired, we don't really have to go to the top. I'm not very good at hiking." What I wanted to say was that hiking combined two elements that humans have had difficulty conquering, insects and gravity.

"Who said anything about hiking? Together we're over 150 years old. We're much too old to hike a whole mountain." The ajeoshi laughed, but looked at the ladies' faces and stopped. "OK, OK together we're, uh, 120 years old."

"We're·going·to·take·a·cable·car·up·to·the·top·hopefully·there·won't·be·a·line·I·swear·everytime·I've·been·here·all·of·Korea's·been·here·waiting·in·line·to·go·up·to·the·top."

We waited in line for only a few minutes; I guess most of Korea decided to stay at home. Cable cars make me nervous. They seem so crude with exposed gears and moving cables. Whenever I go on one, all I do is think about how the cable might break and how the metal lunch box carrying infants, pregnant women, the elderly, and me might plummet toward the rocks below creating an unbelievable carnage that would take rescue workers and search dogs months to reach. Any survivors would have to gnaw off their own arms and legs that are caught in splintered cable and buckled panels of metal and glass. Then come the bears and the fire ants.

"Oh·look·at·the·view·it's·gorgeous·this·little·country·of·ours·is·so·beautiful. What's·wrong·Annie? You·don't·like·the·view?"

I shook my head and looked around the cable car to figure out which area could be safest. If I sat near the windows I could make a quick escape right before impact, but then again, if I sat right in the middle my metal lunch box could protect me from other falling lunch boxes.

"Anne, come here, look at view!" My mother stuck her head as close to the glass as possible without messing up her make-up.

"Get away from the door, Mom. It could open."

My mother rolled her eyes. "How it open?"

"We're in here right? So it can open."

"Anne, come here!"

"I feel nauseous."

"You don't like heights?" The ajeoshi patted my back.

"No I don't like cable cars."

My mother reached in her purse and offered me a hard candy. "Eat, make you feel better."

"What kind is it?"

"Ginseng."

"Barf."

"What does 'barf' mean?" the ajeoshi asked my mother.

When we finally reached the top of the mountain, I scrambled off the cable car and planted my feet on the ground. Beautiful, hard, stable ground. With insects.

"You OK, Anne? You scaredy-cat."

"I'm fine. And cold."

My mother tightened the scarf around my neck. "You such baby."

"The pine tree belongs to Korea." The ajeoshi took a deep breath. "Smells fresh, right?"

I took a deep breath, something I rarely do. Cities smell like piss and people and I usually try to refrain from breathing. The crisp air smelled like damp pine and Earth with a hint of salt from the ocean. Pleasant. We walked up a well-tread path, along with several families, monks, and tourists, and looked out as Korea and the ocean stretched below us. When I was in elementary school, I learned that the sea on the east side of Korea was called the Sea of Japan. My parents were offended and informed me that it should be referred to as the East Sea. The sea belongs to no one, they explained.

After awhile, we all decided it was much too cold to be outdoors, much less on top of a mountain. It took us over seven hours and five meals to get to the mountain and we only stayed there for fifteen minutes. We made our way back down and I survived another cable car ride, but the question was could I survive the car ride back to Seoul?

Sticking to our schedule, we stopped an hour later for coffee and pastries, and an hour after that we stopped at a flea market, and an hour after that we had dinner. The ajuma poured everyone some tea and the adults chatted while I faded in and out of the conversation. I was tired. Traveling for over twelve hours with my mother was no easy task. I'm sure she felt the same way about me.

"You-know-what-we-should-do-now? Karaoke-doesn't-that-sound-like-fun-we-can-sing-or-maybe-we-can-go-dancing-or-maybe-we-can-do-both-I-hear-Americans-don't-like-karaoke-do-you-like-karaoke-Annie-you-know-your-mother-is-quite-a-singer-she-should-be-famous. Let's go!" She tugged on my sleeve.

"No, I'm too tired."

"Nonsense! You always go out all night with your friends; you should be able to go out with us," my mother retorted in Korean. She wagged her finger at me.

"Come on, let's go." The ajeoshi called the waitress over for the check.

"But I don't know any Korean songs."

"They have American songs in those machines now don't you know they have every song in there. It's magic it's from the future. What would Korea do without karaoke? Read?"

I groaned. I am not a big fan of karaoke. The few times I had done it, I drank enough liquor to dull any sense of self-consciousness. My mother doesn't drink, though she could probably benefit from one or five.

Ten minutes later, we were in a private karaoke room where my mother and her friends perused a heavy book of song titles, all in Korean. They ordered tea, queued up a few songs, and let it rip. The background video footage featured young couples in peg-legged acid-wash jeans and frilly blouses. And that was on the men. Each video was more or less the same, a couple walking hand-in-hand on a beach, bridge, or a dock, while lyrics scrolled over their smiling, lovestruck, or pensive faces. I tried to follow along with the lyrics, but couldn't read them before they disappeared off the screen.

"ANNE, SING! SING!" My mother yelled over a pop tune that sounded just like an ice cream truck colliding with a guitar.

"I DON'T KNOW THE SONG!"

"IT NOT MATTER. YOU WORRY TOO MUCH!"

"I can't even read the words!"

"WHAT?"

"I SAID, IT'S TOO FAST. I CAN'T READ THE WORDS."

"JUST MAKE UP WORD."

The ajuma handed the microphone to me and I mumbled a few words. I'm pretty sure I sang something about eggs. The ajeoshi and ajuma laughed hysterically. My mother waved her arms wildly. "YOU HAVE TO SING WITH FEELING!"

"But I don't know what I'm supposed to be feeling!" I passed the microphone to the ajeoshi who belted out the chorus and swayed his head from side to side.

My mother pushed the songbook toward me. She flipped to the American section and pointed, all while singing loudly into another microphone.

"I don't want to sing."

"Anne, why not?" She talked into the microphone so her friends could hear.

"Because I can't sing. I suck."

"So who care? It karaoke. It better if you suck."

"Annie you've got to sing something you can't just sit there you have to have fun."

"But I really don't like to sing."

"You're in Korea. And in Korea people go to karaoke. And when people go to karaoke, people sing," the ajeoshi replied into his microphone. I notice that when people have microphones they use every opportunity to speak into it.

"Anne, just try, go find song. Make Mommy happy for one time. *Ayoo!*"

I grudgingly took the book. The choices available were a massacre of good taste. Most of the popular artists were absent from the book; they probably didn't allow their songs to be butchered into karaoke instrumentals. I tried to find a song everyone would know, which was a major challenge given the selection. Bay City Rollers? Styx? The Captain and Tenille? I shuffled to the P section to look for Elvis. Nothing. What kind of karaoke establishment has Mr. Mister but not the King? I thumbed to the B section to look for the Beatles, the common denominator of music of everyone on Earth. I hear that plants even like the Beatles. The karaoke joint had only one song, "Hey Jude," which is out of my vocal range. Actually all songs are out of my vocal range. I pointed to it and my mother queued it up.

When my song came on, the ajuma passed the microphones to my mother and me for a duet. My mother stood up and straightened her blouse. I remained seated.

"Get up."

"Why? No one else got up when they sang."

"We sing better if we stand. Like at church."

"Nothing can help me sing better except not singing."

"Get up." She grabbed my arm and pulled me up. "You make life very difficult for youself."

A couple on the background video started walking along the beach, and we started singing. I sang softly, allowing my mother's booming vibrato voice take the lead.

"LOUDER! LOUDER!" My mother's friends chanted and laughed.

By the second line of the song, my mother and I were already out of sync, creating an awkward, off-tune echo.

"Second part for you Anne, go!"

I held the microphone limply to my mouth for an awkward solo. I sang a few lines and then stopped. "You know now that I'm reading the lyrics, I don't really get it." I too fell into the trap of taking advantage of the microphone.

"Third part for Mommy. Hey, Jude, don't carry world shoulder, well don't you know fool who play cool. . . . " She swayed her head back and forth just like Stevie Wonder.

Now I realized why we had been out of sync. I started laughing. "OK close enough."

We finished off another verse together and then the ajuma and ajeoshi joined us for the rest of the song.

"You know when I was young, I loved the Beatles so much. I liked Paul the best. I wished he was my boyfriend, but instead I got Annie's father."

"You·did·I·really·liked·John·but·then·he·got·Yoko·so·it·was· too·late·for·me·well·at·least·we·know·he·liked·Asian·women."

The ajeoshi laughed and then looked at me. "Which Beatle do you like?"

To be frank, I am not a Beatles fan. I think they are mostly over-rated but when I say that publicly, people think I'm a baby-eating fascist. "George. I like George."

"Really," he replied. "I didn't know girls liked George. He wrote 'Here Comes the Sun.' He's vegetarian, just like you."

"I thought Paul is vegetarian."

The ajeoshi shrugged and turned his attention to the next song. I tried to follow along for a few more Korean songs with little suc-cess. My mother and her friends picked "easy" songs so I could sing with them, but I still couldn't keep up with the lyrics. After our hour and a half was up in the room, it was time to hit the road again.

As we walked out to the car, the sun had disappeared behind the mountains and the temperature had dropped significantly. My mother dug in her purse and brought out a bag of roasted chestnuts left over from this morning. They were cold but everyone took a chestnut anyway.

"You carry so much crap with you."

"It not called crap if you use."

In an unprecedented move, the ajeoshi announced that we would go the rest of the way—nearly two hours—without stop-ping. It was midnight, time to get home. The ajeoshi cranked up the heat and the conversation became slower and quieter inside the warm comforting womb of the car. The ajuma was the first to drift off to sleep, followed by my mother. I considered sticking my finger in her ear and poking her in the ribs, but for the first time that day, it was quiet. I peeled a chestnut, leaned my head back, and chewed in silence.

RULES OF ENGAGEMENT

I was drifting off to sleep. My father's car pulsed slowly in traffic and I found the gentle stop and go soothing.

"You want more air con?" My father waved his hand in front of the vents. "I can make more cool."

"No, I'm fine." I leaned my head back and looked out the window. In true L.A. fashion, one driver was closing a deal on his headset while creeping his SUV into the small space in front of us. My eyes struggled under the weight of a late night and a bright morning sun. My father reached over and put the sun visor down for me.

"You can put seat back."

"I'm fine."

"But you be more comfortable."

"No, no, I'm fine."

"There so much traffic. So much." He shook his head.

I looked at the clock. "We have plenty of time." I yawned. Morning flights back to New York are always rough.

"I guess everyone go LAX." He chuckled. Whenever we're stuck in gridlock, my parents joke that everyone is headed to the same place as us. If we're on our way home, my mother says that everyone is going to pile into our house and expect food and beverages. I shifted in my seat and closed my eyes.

"Very hot today, Anne, very hot. You can see how hot! Look!"

I kept my eyes closed. "Yeah, I know, I can see it. It's all very hot."

"But it so *hot*."

"It's summer."

"You want more air con?"

"I'm fine, Dad, really. Now it's sleepytime for Annie." I turned toward my window and curled up for a nap on the 405. Kenny G. came on the radio, but I was too tired to stick a fork in my ear or change the station. My father likes easy listening, the easier the better. His favorite American song is "Lady in Red" by Chris de Burgh. I once found a yellow notepad with the lyrics printed in his neat chemist's handwriting. He was preparing to woo the pants off everyone at karaoke. I dozed off to the soprano sax mastery that is "Songbird." For about five seconds.

"Anne?"

"Mmm?"

"You sleep?"

"Yes."

"You tire?"

"Yes."

"Hot?"

"No."

"Tire?"

"You already asked me. Yes. Tired. Sleepy."

"Anne?"

I sighed and turned toward him. "What?"

"I'm an old man."

I groaned. My father, like many fathers, is bad at starting conversations. He always starts with something nonsensical and as he fumbles toward a point, I have to piece it all together. It's like figuring out what happens in a story by reading every other page. He also has bad timing and will wait for the worst moment to have a serious talk: right before I'm rushing out the door, in the middle of Best Buy the day before Christmas, as I'm falling asleep. Most of the time, our conversations are about things I already know, like the importance of a good education and a stable job. Sometimes, he doesn't have the conversations we're supposed to have, say when someone in our family has cancer.

"What are you talking about?"

"I'm old, you know?"

He was sixty-three, but his face only looked fifty-five except for his mouth, which until recently looked about one hundred years old. Over the last few years, my father lost most of his teeth because, according to him, he had bad teeth. I tried to explain that they were bad because he never flossed, but he explained that it was genetic because his parents also had bad teeth (I'm pretty sure my grandparents never flossed either). Last year he took out a fat personal loan and got jawbone grafts and titanium implants. I told him that he better not die before paying off the debt or else I'd rip his fancy teeth right out of his corpse and use them when I started losing my teeth because, you know, it's genetic.

Traffic on the 405 had come to a complete standstill and hundreds of brake lights dotted the freeway in front of us. I abandoned all hope for a nap and sighed. "Why are you an old man?"

"Because I'm not young."

"Most people aren't young."

"You right. You not young." He shook his head in disapproval. "Not young."

"No, I'm not."

"You old. Not as old as Daddy, but you old."

I rolled my eyes. "What's your point?"

"Soon you be thirty-seven."

"What?" I was twenty-eight.

"You be thirty-seven."

"In like ten years."

"Nine year."

"Whatever."

"You going to marry?"

"Are you kidding me? We're not talking about this. Again."

Until I finished college, my parents had never taken an interest in my love life. In fact, it was quite the opposite. "Anne," they'd tell me, "school not for fun, only for study to be doctor." In high school, between the track-and-field practice, flute and piano lessons, SAT classes, and tutoring sessions, there was no time for boys, not that any boy would even desire four feet, ten inches of me. In college, when I grew four inches and a rack and was finally able to pursue the elusive male, my parents made me promise to concentrate on my studies, and I told them not to worry—that if there was one thing Berkeley had, it was a lot of things to study (for example, physics, the major with most guys, and architecture, the major with the best-looking guys). When my mother asked with raised eyebrows if I "see boy at school," I said of course not, because that was actually the truth. It turned out that all the eligible guys at Berkeley had found other things to study. Shortly after I graduated college, when it was finally acceptable for me to date, my parents began asking why I hadn't found a husband, and they haven't stopped since. Each year they get more desperate. I'm pretty sure they've mounted my biological clock on the wall, next to the picture of The Last Supper on petrified wood.

"Annie, why you not marry? You have to marry now."

"Now? Right this second?"

"I'm an old man. I get sick."

"What are you talking about? You're not sick."

"But I get sick someday. Maybe tomorrow, maybe in five years, who know?"

"You're not sick. You're insane."

"Annie, you have to marry before I get sick."

"I'm not getting married just because you might get sick someday. That is stupid."

"You know, man who is thirty-seven look for twenty-four year old."

"What are you talking about? Everyone's looking for a twenty-four year old. Who's *not* looking for a twenty-four year old?"

"But you not twenty-four. Everyone who thirty-seven not look for you."

"What is your obsession with thirty-seven?"

"When you thirty-seven, no man want to marry you because they want younger."

"*Dad,* I'm only twenty-eight! I'm not old! I'm closer to twenty-four than I am to thirty-seven." I looked desperately at the freeway. It was still packed with cars, bumper to bumper. I considered getting out and walking, but my father was right—it was very hot out there. Waves of heat were rising from the asphalt, as if Hades were bubbling just underneath the San Diego Freeway.

"You think that now, but you get old very fast. Women get old very fast."

"You know what else is getting old very fast? This conversation. Yoon-chong isn't married. Andy isn't either. Or Mike. Or Woo-jay or Tina. And they're older."

"They in big trouble, Annie. Big trouble. They not find husband or wife now, they never will."

"That's ridiculous. How do you know they haven't found someone already? I'm sure they're all dating someone serious; we just don't know it."

My cousins and I try to maintain a separation between family and people who are important to us. Whenever we get together during the holidays, my cousins ask me if I'm dating anyone, and I always answer no. Then I ask them the same question so that they can give the same answer, and that's the end of that conversation. We do this not because we aren't interested in keeping up with one another but because in a family as large as ours someone is always listening. And in a family as loud as ours, someone is always talking. And if someone is talking, they're probably talking trash. Though we hate to admit it, we care what our family thinks; we've been brainwashed to seek approval and obey, just like the rest of Korea's children. As a result, the unspoken rule among the cousins is that we only discuss our significant others when it's absolutely necessary. The first time many of us learned about Yoonmi's boyfriend was when they were engaged. When she brought her fiancé to a big family dinner, all the cousins observed closely because she was forging new territory, setting a precedent for what would happen when an outsider infiltrated family lines. The cousins watched with fascination, fear, and pain as Yoonmi's fiancé answered tough interview questions: Where did you go to school in Korea? (Yonsei University, which is the Yale, not Harvard, of Korea.) What do you do for a living in the states? (lawyer) How will you provide for "our Yoonmi?" (The best he could—by working hard, making sure she was happy.) How many children do you want? (Two or more, depending on how Yoonmi felt.) Where do you want to raise the family? (In Los Angeles, close to Yoonmi's family.) Where do you want your children to go to college, and what kind of professions should they have? (Harvard, lawyer or doctor of course.) Would you consider raising the family in Korea? Why or why not? (Yes, but his practice is here and he'd have to check with Yoonmi.) He was smooth—accommodating but not obsequious, eloquent yet warm and friendly. In the end, everyone liked him because he looked smart, had nice hair, and was a well-educated lawyer from Korea who obviously worked well under duress. Actually he is pretty perfect now that I think about it; fluent in Korean and English, but definitely more Korean than American, and the owner of a luxury German sedan. Still, the whole ordeal was so stressful that the cousins only wanted to go through it once. It didn't matter how serious our boyfriends and girlfriends were; there was no need to bring home or even mention anyone except for a fiancé. But there was one cousin who ignored the memo. Twice.

Four years my senior, Andy works in Los Angeles as a physical therapist. He's a good-looking mild-mannered guy with a lucrative career, and we all figured he'd be next to marry. In 2001, he brought a girl named Eunice to our family's New Year's dinner. They had been dating for four years, so marriage was a definite possibility, but they weren't engaged. I wasn't at the dinner, but from what my mother said I gathered that Eunice was very smart, very nice, and very Filipina. Since English was a challenge for most of our relatives, Andy answered most of the questions and translated for his girlfriend. The family wasn't thrilled with Filipina Eunice, but there wasn't much they could do. At least she was smart and nice. They got over their disappointment, or at least pretended to any-way. Andy and Eunice didn't work out, and in 2003 he brought another girl to the family's New Year's dinner. This time her name was Julie, and she was an elementary school teacher, and she was Chinese. Throughout most of the night, my relatives left her alone. There was a feeble Q&A session, but most decided it wasn't worth getting to know her, not if Andy was going to bring another girl the next year anyway.

"I can't believe. Why Andy bring new girl?" My mother threw her hands up in frustration. My mother, father, and I were driving home after New Year's dinner. My brother had gone off in his own car, back to his apartment on the other side of the San Fernando Valley. Lucky him.

"What are you talking about? I liked Julie. She was really sweet."

"He bring different girl every year." My father shook his head. "Very bad. Why he do that? Why his mom not say anything?" Andy's mother is my father's youngest sister.

"He does *not* bring a different girl every year. He's only brought two in his entire life."

"Two is too many." My mother turned her head to look at me in the backseat. "Only need one. Two make people confuse."

"What's so confusing? There are two girls: he dated one; they broke up; and now he's dating another. It's actually all very easy if you pay attention."

"He act like a fool." My father gripped the steering wheel and concentrated on the road. "Andy should only bring most important girl. He should know that."

"Well he thought Eunice was the most important girl."

"Filipino." My mother added.

"Yes, he thought Filipina Eunice was the most important girl. And now he thinks Julie is the most important girl."

"Chinese."

"Whatever. I heard that in other families, everyone meets boyfriends and girlfriends all the time." I wasn't sure why I was defending Andy—I didn't really want to expose every boyfriend I'd ever have—but I think I was defending reason and common sense, two things my family could use, along with patience and a volume button. "Andy and Julie have been together for over two years. Maybe they'll marry."

"Maybe they not marry," my mother contended, "Who know? That why so confuse."

"But it's not confusing!"

"Anne, you not understand." My father shook his head.

"We only interest in one you marry," my mother added.

"Most important one," my father echoed.

"I know. I get it, but it's not fair for you to get mad at Andy just because he brought a girlfriend to dinner. Don't you care who he's dating? Because you're supposed to care."

"No, we care. . . . . We just don't want to know." My mother adjusted her seatbelt so she could turn around even more and look me squarely in the eyes. "When you find husband, you tell us. Don't bring home everyone you meet. Only bring one."

"Don't be like Andy," my father warned me.

Throughout my life, my parents urged my brother and me to be more like Andy: be more studious like Andy; be better at sports like Andy; speak and write Korean and go to church every week like Andy. Now they were saying, just kidding, forget Andy. Don't be like him.

"OK, so you want me to tell you when I've found a husband, but I shouldn't tell you who I'm dating? That's confusing."

My mother rolled her eyes. "Why you ask so many question? I thought you say not confusing! *Ayoo*, I give up." She looked at my father. He shrugged.

"Don't worry," he told her in Korean, "she'll figure it out. She's smart. Someone will marry her."

"You think?" my mother asked him in Korean. They laughed.

"I speak Korean, too, you know."

"You don't speak Korean very good, Anne." My mother smiled impishly.

"You mean 'I don't speak Korean very *well*.'"

"*Ayoo*, Anne, you never find husband because you cause so many problem."

"What did I do? I'm just sitting here. No problems here."

"No problem?" my mother asked, "So that mean you find husband? Where is he?" She looked around the car.

"Why do I have to find the husband? Why doesn't the husband find me instead?"

"You need help? You know Dr. Kim's son, Daniel, he orthodontist—"

"MOM, STOP!"

The car was silent. Finally.

"I think Daniel engaged already," my father whispered.

"DAD!"

"See, Anne? You run out of time!"

❈   ❈   ❈

My parents want the impossible. They want me to get married immediately, but to the "perfect" man who apparently is waiting for me to find him. Where do I even begin? Whenever I want to find anything, I consult a map, so I typed in "Annie's husband" into Google Maps and surprisingly it came up with four pages of options, including a bed and breakfast in Springville, California; an outlet store in Lenox, Massachussets; and a nanny referral service in Mountain Village, Colorado. Sadly, none of these places deliver to New York, but it's comforting to know you can find anything on the Internet. The pressure my parents put on me to find a husband makes me a bad girlfriend, but it's not because I feel the need to get married in the next five minutes. In fact, I'm in no rush to get married. Unlike my parents, I can't hear my b-clock ticking away. In fact, my clock could be in its final dark and lonely hour and at this very second, my withering ovaries could be spitting out the last tired egg, but I wouldn't know it. That is not the kind of pressure I feel. Instead, I feel the same pressure I felt when I brought home my first B+. I know that when my parents ask me if I've found a husband, they are referring to a very specific kind, a Korean, Catholic, Harvard-educated doctor/lawyer/orthodontist husband. My last few boyfriends have not fulfilled any of these requirements. The one who came closest was Aaron, who only went to M.I.T., which is down the street from Harvard, but he was white and Jewish. Korean Catholic men from Harvard just don't interest me. The problem is, if I marry a man I like, my parents are in for a big let-down, and as much as I hate to admit it, I don't want to disappoint them. So it seems easier to never fully commit to anyone than to commit to someone and disappoint my parents. This makes me a bad girlfriend.

I dated Aaron for six years and even lived with him at one point, and I never told my parents about him. On the other hand, I didn't exactly hide Aaron from my parents either. I've known him since I was twelve and we both attended the same junior high, high school, and college, and even moved to New York together. My parents knew him as one of my best friends. Aaron was the brightest kid in high school, but he also dyed his unruly blond hair and his King Tut goatee blue. When we both returned to Los Angeles for the holidays, I always invited him over to my parents' house for a Korean dinner, as I did other friends. He always arrived on time, with fruit (the appropriate Korean dinner gift) or my mother's favorite coffee beans (Swiss Chocolate Raspberry) in hand. He did everything right, with only a little coaching from me. Aaron and my parents, especially my mother, got along well. He was charming at the dinner table—making jokes, complimenting my mother's cooking, and shooing her out of the kitchen so he could do the dishes. Since he was always around, I figured my parents would guess that we were more than friends, and I hoped my mother would be smitten with Aaron and say, "Anne, why you not marry Aaron, he seem like such nice, Jewish boy." But she never did. Of course, I could have said, "Mom, Aaron is a nice, Jewish boy, I like him. Why don't I marry him?" But I never did because I knew how that conversation would go:

"OH MY GOD. Such shame! Such shame!"

"But he's a nice guy. He's smart and funny and he smells good most of the time."

"Why you do this to you Mommy?"

"I'm not doing anything to you!"

"Yes you do. You give me cancer. In my heart." She clutches her chest and weeps, her eyeliner flowing down her face in black rivers. "How I survive this? My only daughter marry JEW!" She shakes her spatula at me and a little hot oil splatters on my clothes. She has been frying scallion pancakes.

"He's not orthodox or anything. He eats pork."

"Then he bad Jew! You can't even marry good Jew! How he eat kim chee?"

"What are you talking about? He loves kim chee. He'd eat a lobster stuffed with kim chee and bacon if you let him."

She looks around the room, her eyes spinning wildly in her head, and runs to a window. "I jump. If you marry him, I jump!"

"You're not gonna jump, stop being crazy."

"Yes, I jump." She slides the window open. An Arctic gale gusts through the room. It appears that hell has frozen over.

"No, you won't."

"Yes. When I die you say, 'Oh I such bad daughter. Waaah!' You come to Mommy funeral but you have to sit in back of church." She looks at me with tears in her eyes, and looks out the window.

"Don't do it! You're being ridiculous."

"How come you love Jewish Aaron more than you mommy?" Then, she jumps. I roll my eyes because our kitchen is on the ground floor.

Aaron never pressured me about marriage; he knew about the unspoken family rule. Any motions toward matching jewelry had to come from me. But even after six years of dating, I was still scared to let our relationship go any further. I fooled myself into thinking it was because I didn't love him, but I think knowing that I was going to disappoint my parents prevented me from being fully committed. When Aaron and I broke up, I was sad, but relieved. It was as if I had averted a conflict that was six years in the making.

Recently, my cousin Andy came to visit a friend in New York for a weekend, and we met up for brunch. I hadn't seen him since

2003, when he brought Chinese Julie to New Year's dinner. As always, he looked healthy and happy, though his facial hair was a little disturbing.

"What's that strapped to your face? A hamster?"

He stroked his chin. "What, you don't like it?"

"I don't like hamsters. They're like mice that you keep around on purpose."

He laughed and we walked toward a café in the West Village. His eyes tracked all the trendy New York girls as they strutted down the street carrying shopping bags and towing little dogs wearing sweaters.

"You like what you see? You think they're better looking in New York?"

Andy grinned sheepishly. "Actually, I was looking at the way they walk."

"Oh sure, I look at the way guys walk all the time."

"No seriously, it's what I do." Andy is a physical therapist and he was recently promoted at his clinic. "Half the time I don't even look at them, I'm just looking at how messed up their knees are, and then I think of exercises they could do to fix them." I laughed, surprised by my cousin's geekiness, normally he is a smooth-talking, slick guy.

"It's the pointy shoes. Those things are deadly. If you opened a practice here you'd make a killing just from pointy shoes."

"I'm working too much. This is the first vacation I've taken in four years. I never have any time."

"What about Julie? Are you guys still together?"

"No, we broke up."

"Sorry." I could hear my parents wail, *See? So confuse!*

"She went off to pharmacy school. Couldn't handle the distance."

"I liked her, she was cool." I shrugged. "You dating anyone now?"

He shook his head. "No, too busy. You? You got anyone to bring home to your parents?"

I paused. There were no parents or uncles or aunts around. It was just us. I could tell himabout every guy I've ever dated, but I decided to be cautious. His mother, like mine, talks a lot. Maybe too much. "Well, you know, I can only bring home the guy I'm going to marry, so that's not happening anytime soon."

He laughed. "It's the same in my family too. Like I never told my parents who I was dating, and then I was talking to my mom one day and she was like, I feel that I don't really know you, like I know nothing about your personal life. So I was like, yeah, you're right, we should be closer and all that stuff. So I started bringing around my girlfriends."

"And bringing them to family dinners."

"Yeah, and then one day my mom was like, who are all these girls? Why do you keep bringing all these girls home?"

I started laughing. "Because she told you to!"

"No kidding. She was so confused. So then I was like, I thought you wanted to be close and know more about my personal life. And she said, actually no. Don't bring anymore girls around until you get engaged."

I laughed; it all made sense to me, why my cousin introduced two girls to the family. But I did have one question. "Why didn't your mom stop you from bringing your girlfriends to meet the whole family? You know, my parents were pretty annoyed. I'm sure the other uncles and aunts were too."

Andy stopped laughing. "Really?

"Yeah."

He shook his head. "She didn't say anything. I thought that was what my mom wanted."

"It *was* what she wanted and then it wasn't, I guess."

Andy shook his head. "I've got to stop listening to my parents."

"No, just hide everything. It's healthier that way."

We finished our brunch and walked back to my neighborhood. I pointed out where to get the best pizza in New York (John's on Bleecker Street) and the most overrated cupcakes in the city (Magnolia Bakery), where he could buy expensive "yoga-inspired jewelry" (Satya), and I pointed to a salon that specialized in Scandinavian hair designs, whatever that meant. When we reached my block, I pointed to a store called LeSportsac.

"Contrary to popular belief, they don't sell jockstraps. They sell handbags."

He grinned. "Would you ever move back to L.A.?"

"Maybe. I don't know. Our family's annoying. But I feel closer to them by being far away. Does that make sense?"

"Kind of." Andy laughed. "Hey, you know, I think this is the first time we've ever hung out without our families."

"No kidding?" I thought for a second. He was right. In twenty-eight years, I had never spent time with Andy alone, even though we grew up twenty minutes from each other. "You know, you're pretty cool. I'd actually hang out with you even if we weren't related."

He smiled, took out his digital camera, and took a picture of us in front of the overpriced salty French bistro on my corner.

One of the advantages of being the youngest in my family, or at least in my generation, is that I can watch while my older cousins forge new ground and I can learn from their mistakes. Yoonmi and her fiancé taught me that no matter how perfect the guy seems and no matter how happy the guy makes me, he'll still be under a lot of scrutiny from my family, and chances are, if he's dating me, he'll have a lot of faults. Andy and his girlfriends showed me that I shouldn't expect a warm welcome for someone whose name is Filipina Eunice or Chinese Julie or Jewish Aaron. In the end, I am a coward. I date guys who are good enough for me, but I question whether or not they'll be good enough for my family, and this holds my relationships back—not only my intimate relationships but also my family relationships. I could be like Andy and bring someone important home; someone whom I may or may not marry. But my fear of disapproval is hard to overcome. How can I be truly committed to my boyfriends if they can't meet the people who are important to me? And how can I be truly close to my family if I have to hide someone important from them? Maybe I just have to take a risk and believe that the guy I love will be someone my family will love too.

new year's games

"So, will everyone be around for Christmas this year?"

I cradled the phone to my ear while I searched the Internet for flights to Los Angeles. As usual, my mother was in the car and the radio was blaring in the background. This time a Korean vocalist was singing a traditional folk song; her voice wailed and cracked and made my toes curl. Korean folk singers always sound drunk and clinically depressed.

"Yes, Anne, we be here."

"You're not going to Las Vegas? You're not going on some golf trip or something?"

Last Christmas, I told my mother I couldn't come home for the holidays because of work. She got so upset that I convinced my boss to let me take a working holiday. This thrilled my mother and she chattered gleefully about cooking an elaborate Christmas dinner for the family—turkey and the trimmings plus a Korean feast. I arrived on Christmas morning and discovered that my mother had gone to Las Vegas with her sister and mother. My brother decided to spend Christmas with his friend in San Diego, and my father was tied up with a big project at his lab. I was infuriated and vowed never to come home again, not until my family had been replaced by a real one.

"Anne, I tell you, we stay here."

"And for New Year's Day? That's still going on? Everyone will be there for that?"

"Yes, we go you uncle house."

New Year's Day is an important holiday for Koreans. Our entire extended family gathers together for a party, during which we point out one anothers' weaknesses. I'm too short, my brother is too fat, my aunt is too loud, my older cousin Yoon-chong is too old to be single, etc.

"OK and you *sure* everyone's not gonna take off somewhere without telling me? You're absolutely, positively sure? Like one hundred percent sure?"

"ANNE. YES."

"So, let me make this clear, EVERYONE will be there?"

"YES, EVERYONE. OK now you stop bother Mommy. You make me such headache and you not even here yet!"

Right after I clicked "Confirm Your Order" on the United Airlines webpage, I felt anxious. Seven days with my family by my own accord. I must be high on crack. How would I survive? How would we survive each other? Is pepper spraying a loved one legal in California? I scratched my neck until it was bright pink. Family has always made me a little rashy, a physical manifestation of emotional irritation.

The flight from New York to Los Angeles was heinous. I sat next to a chatty obese woman who had to be shoehorned into her seat. Her excess invaded my space and I had to wedge my hand between her folds to find the end of my seatbelt. Her necklace had enormous silver disks that I swear were the size of dinner plates. She kept on asking me questions while I was watching the in-flight movie or reading my book or sleeping. She was getting on my last nerve, which I had been saving for my family. I missed my connecting flight in Denver and was stranded for several hours. When I finally arrived at my parents' house, completely exhausted, I found out Mike wasn't coming home for the holidays. I felt cheated. Apparently, when my mother said "everyone," she meant "everyone except my brother."

"Mom, you told me everyone would be here. Why didn't he come? I came. He should come. He has to suffer with the rest of us."

My mother was getting ready for her church's Christmas party. Just as I did when I was little, I leaned against the bathroom wall and watched her meticulously apply makeup in the mirror. When I was young, I knew exactly which item she needed next and handed it to her—loose powder, a palette of eye shadow, lip liner. She took the mascara brush away from her lashes so she could roll her eyes at me.

"Anne, what you mean 'suffer'?"

"I mean I'm here, so Mike should be here."

"He just move. So I tell him not come."

"What? Why would you do that? Remember last year? I told you I couldn't come and you got all upset. You were like, 'Ohhh my only daughter not love me waaah.'"

I threw my head back melodramatically and slapped the back of my hand to my forehead, like a dying Shakespearean heroine or a Southern belle with a case of the vapors. Then, remembering the misery of last year's lonesome Christmas, I scrunched up my face and looked cross. My mother laughed.

"Oh, Anne you such comedy! Don't make ugly face, you get wrinkle!"

I grimaced even more, making deep crevasses in my forehead. My mother grinned and handed me a small jar.

"What's this?"

"Wrinkle cream. Put on you face. Why you face so dry?"

"What are you talking about? My face is fine."

I looked in the mirror and saw patches of white flaky skin on my cheeks. I was molting. I sighed and opened the jar and dipped in my index finger. It looked like whipped cream cheese but smelled like flowers.

"You're not gonna make Mike come out here?"

I rubbed the cream cheese vigorously into my face. My mother watched me and cringed.

"No Anne, make small circle, small circle. Be gent. *Ayoo*, you have so many wrinkle."

"No I don't."

"Yes you do."

"No I don't."

"Yes you do. You look like Grandma."

"No, I don't. I look like Annie. What are we, five years old? We're not talking about wrinkles here, we're talking about Mike."

"Anne, he leave L.A. two week ago and move Chicago. So silly for him come back right away. But I so happy you come back and see you mommy! Who cut you hair? Why so short? Are you boy or girl?"

I groaned and fled the bathroom before my mother could corner me with her curling iron and a gigantic can of hair spray.

I had wanted my family to be together for the holidays, even though they make me grind my teeth into little nubs. In the end, however, we are family and we should spend time together, even if it kills us. But now my brother wasn't coming and I felt the dread of New Year's Day dinner at my uncle's house. I started to regret coming back. Why did I want this?

My brother and I aren't the best of friends, but we are allies. Mike makes New Year's Day dinner with the relatives more palatable. We sneak quiet jokes and exchange knowing glances across the table when relatives start bickering. When we can't understand the conversations in Korean around us, we commiserate over our soul-sucking bosses or discuss how Eddie Murphy's career ended up in the toilet. I didn't want to do New Year's with the family alone.

"Mike, are you *sure* you don't want to come out? It'd be totally awesome!"

"Are you out of your mind?"

"But there'll be wontons. Hot, delicious, savory wontons filled with . . . stuff I don't eat, something mammal. Think about wontons—you love them. Do it for the wontons."

My brother loves food and he is shaped like a hippopotamus. I've learned to never stand between Mike and the kitchen. If my brother wouldn't come to L.A. for the company, maybe he'd come for the food.

"Sorry, Anne, but fuck that."

"Fuck wontons?"

"No, fuck going out there. I just moved here, dude. It'll be a pain in the ass to go back."

"But there'll be tasty, fried wontons . . . mmm wontons . . . extra fried . . . extra tasty. . . . " I made chomping and slurping noises over the phone.

"Anne, Jesus, what the hell? Stop being such a little shit."

"Don't make me go there by myself. It'll be horrible."

"Dude, there's no way I'm finding a plane ticket now. Besides, it won't be that bad. Mom and Dad will be there." His throaty laugh was tainted with the kind of evil practiced only by big brothers and tobacco companies.

"Oh, so cold, Mike. So cold."

"Whatever, just keep your shit together and smile. Easy. And hey, merry Christmas and happy New Year, you little bitch. Drop me some e-mail."

"Same to you, jerkface."

I hung up the phone. On New Year's Day I would be on my own. All I had to do was keep my shit together. And smile.

On the first day of the year, I woke up at three o'clock in the afternoon with a mild hangover and a serious need for coffee. I stumbled into the kitchen, my eyes swollen and crusty and my head in a brutal pre-caffeine haze. I peered at the half-pound bags of coffee beans in disgust. My mother doesn't believe in coffee-flavored coffee. While I deliberated in agony between Supreme Holiday Pumpkin Cinnamon Cardamom Blend or Swiss Chocolate Raspberry Hazelnut Awakening, my father walked into the kitchen. He had on a red collared shirt and a yellow sweater, a gift I gave him last Christmas. He had on no pants. He did, however, have on underwear.

"Jesus Christ, DAD, put on some pants!"

"I can't find any pants. I think they all dirty."

"How could they all be dirty?"

"No pants, I have no pants."

"Did you look in the laundry room? Can you at least wear a robe? Seriously, you're killing me here."

I wrenched my eyes away from his scrawny, pale legs. As he's gotten older, my father has lost weight in his legs, but gained in his belly. When he turns to the side, he looks like the letter P. He returned to the kitchen triumphantly, holding up a pair of blue and green plaid pants. They looked familiar.

"Uh, I think those are Mom's pants."

"Really?"

He looked at the tag and went back to the laundry room, muttering to himself. My mother walked into the kitchen, fully dressed with hair and make-up in place.

"Anne, why you in pajama still?"

"I just woke up."

"Oh my gosh, it three o'clock! I already went church and come back. How you can sleep all day? Oh look you hair!"

I looked at my reflection in the oven. My dark hair was sticking straight out in a million directions. My head looked like a sea urchin.

"I just woke up, give me a break. Tell me, do we have any normal coffee?"

My father joined us in the kitchen, defeated. No pants.

"*Ayoo,*" my mom groaned in Korean, "Have you no shame? Your daughter is standing here. Where are your pants?"

"He can't find any," I answered in English, "Do you have any normal coffee? I don't want pumpkin pie coffee; it's disgusting."

"Anne, you can drink my coffee," my father said, skittering

across the kitchen. As he reached up to open a cabinet, the bottom of his shirt raised to reveal a hole in the seat of his briefs.

"Oh. My. GOD! MY EYES! MY EYES! They burn!"

"*Ayoo,* shame! Shame!" My mother shut her eyes in horror and stamped her feet.

"What? What?" My father looked at us incredulously. "Everyone so LOUD! Stop scream!"

"Anne," my mother cried, "now you know what I see everyday. How I live like this?"

"I don't know, but I'm glad I don't live like this. You people are nuts. Dad, you really should wear pants when you come into the kitchen. It's not sanitary."

My father scowled and handed me a jar of Sanka.

"Anne, you be nice. My coffee taste better than Mommy's."

"Oh no way. This is instant coffee. I don't do instant. No one should ever be in that much of a hurry."

"You complain too much," he told me.

"You're not wearing any pants," I said flatly.

"Everyone so crazy, how I live?" my mother wailed. She looked up at the ceiling, toward God, "How I live?"

"I bet God drinks regular coffee," I grumbled.

"Anne!" my mother cried.

"Where's my pants?" my father demanded.

"Where's the normal coffee?" I demanded. I felt a sharp pinching in my temples and a throbbing between my eyes. I started scratching my arms. I dug my nails deep into my forearms.

"OK, everyone, pay attention," my mother snapped in Korean. She put on her drill sergeant face, the look that once instilled fear in my brother and me throughout our childhood. "We are leaving at five o'clock. Everyone get ready to go. Anne, you can wait until your uncle's house for coffee—STOP SCRATCHING. They'll have a bathtub full of coffee just for you. And you," my mother looked warily at my father, "I don't know how you've lost all your pants. How do you lose pants?"

My mother marched out of the kitchen in search of pants. There was a moment of silence as my father and I listened to her slippers shuffle down the hallway.

"And Dad?"

"Yes?"

"You can't wear that sweater with that shirt. They don't match."

"What you mean? Red and yellow match."

"No it doesn't. You look like you work for McDonald's."

My father laughed and looked down at his sweater. "You bought this for me last Christmas. I like it. Yellow is color for emperor."

"I know, but you aren't an emperor."

"What you mean? I'm emperor of the house!"

"Well then I guess the emperor works at McDonald's. And yes, I would like fries with that."

"Anne?"

"Yes, Dad?"

"You give me headache."

He snickered as he walked out of the kitchen, half naked. Or half dressed, from an optimist's perspective. My bloodshot eyes throbbed. My mother keeps a bottle of Tylenol on the kitchen table, next to the salt and pepper shakers, for easy access. I opened the bottle, only to find it empty. Now I was suffering from severe headache and irony.

I doused my hair with water and three different gels to tame the sea urchin. I threw on some clothes and waited in the kitchen for my parents. I watched the re-run of the Rose Parade. In high school, I used to decorate floats with my friend Janna. I'd come home covered in glue, petals, and seeds, and leave a sticky, fragrant trail straight to my bedroom. This infuriated my mother, though she liked the flowers I'd bring home for her.

"Anne, why you not ready?"

"What are you talking about? I'm ready. I'm waiting for you guys."

She slowly looked at me up and down. One side of her lip curled up in displeasure.

"Mom, come on, I look fine. Leave me alone."

"Why you not wear skirt?"

"Because I didn't bring a skirt. It's cold in New York."

"Why you always wear pants?"

"What are you talking about? You're wearing pants. Dad's wearing pants. Hopefully."

"Can you wear Mommy blouse? I have white one, very pretty."

I realized that I had spent most of the New Year bickering over clothes. I was exhausted and we hadn't even left the house yet.

"Mom, I don't want to. Can we please just go? Please?"

"OK, OK, when you see everyone, make sure you say thank you to Tina Mommy."

"For what?'

"She give you Christmas gift!"

My mother pointed to the kitchen table. There sat the largest jar of peanuts (in the shell) I have ever seen in my life. I couldn't believe I had missed them before. Our ancient kitchen table was struggling to support the weight.

"You've got to be kidding me. What is that, three gallons? Five? What am I supposed to do with this?"

"I know. Who give peanut? How you can eat all?"

"What do you mean how can *I* eat it all? There's no way I'm taking seven tons of peanuts back with me. We'll crash."

"Maybe you share with everyone on plane."

"Mom, I can share this with everyone in New York City and we still couldn't finish it. I don't even like peanuts. Why not pistachios? Almonds?"

"Peanut cheap."

The jar's label had Korean writing on it. Its original contents were for kim chee. My aunt had purchased a gigantic bag of peanuts, probably from a circus supplies store, and filled jars she had around the house. I opened the jar—it smelled like peanuts and cabbage. I gagged.

"*Ayoo*, Anne, just say thank you."

I grumbled at my aunt's thriftiness, though I didn't get her a gift. My father walked down to the kitchen. He had found some pants. He had also changed into a blue shirt. I was going to comment, but decided against it. Our family could use a little more self-censorship.

"All my pants at dry cleaner. So I have to wear old pants," he said, tugging at the waist, "I think maybe they a little tight."

My mother rolled her eyes. Everyone reached for keys off the hook near the garage door.

"Who drive?" my mother asked.

"I'll drive," my father replied.

"I'll drive, too," I added. Silence. My parents raised their eyebrows at each other. My brother always insisted that our family take two cars to my uncle's house so when the moment is right, we could both escape the party together. He used the excuse of working early the next morning, and that was the end of the conversation. But now, I was on "vacation." Plus, it was Saturday. I was trapped.

"I'm going to a friend's house tonight," I began. I looked at their stern faces. "*After* the party, of course."

"You stay for whole party, Anne," my mother warned.

"Until the end," my father added, tugging at his pants.

"Fine, fine, I'll stay until I die, let's just GO."

My uncle's house is a half-hour drive from my parent's house, in a pleasant but flavorless part of the San Fernando Valley. My father's little brother and his family immigrated to the States in 1991, when I was fourteen. They lived with us the first few months they were here—nine people in one house with two bathrooms. It was a nightmare. Once my aunt graciously packed me and my brother lunches for school—peanut butter and ham sandwiches. She didn't quite understand the concept of peanut butter. As I pulled up behind my parents' car at my uncle's house, I realized I should've gotten coffee beforehand. I can be quite cantankerous without caffeine and my aunt probably wouldn't brew anything until after dinner was finished. I couldn't even rely on the presence of soothing booze. For some reason, the Choi clan doesn't drink much at family gatherings, which makes alcohol hard to come by at a time when I need it the most. To toast the New Year, there is usually one bottle of wine for seventeen people. Inexplicably, this wine is usually Manischewitz. My brother and I always joke about it, complaining about how hard life in Israel was for us Korean Jews and asking whether the wontons are kosher. No one else ever appreciates the humor, not even Tina and Andy, who immigrated here when they were young.

My aunt and a heat wave of garlic and fried food overwhelmed us at the door.

"Annie, Annie, you are here! Our Annie has come! Have you gained weight?" my aunt said in Korean, looking me over. She gazed intently at my face.

"No, no, I'm the same. Happy New Year."

I gave her the customary bow. My aunt nodded and shot my mother a grave look.

"I think her breasts have gotten bigger. What's wrong with her skin?"

"I know, it's so dry. You think her breasts are too big?"

"No, no, it's just that she's little, but her breasts are big."

Since my Korean comprehension outshines my verbal skills, I can listen, but I can't react. It's incredibly debilitating and infuriating. But this is probably for the best. Otherwise, I'm sure I'd say something I'd regret.

"I'm the *same*," I hissed.

I see my aunt once a year, but she seemed comfortable enough to talk about my chest. If given the chance, I'm sure she'd feel comfortable discussing the state of my vagina too. I forced a smile and a dry laugh and pinched my mother lightly. She smirked at me and ushered me into the living room.

My brother wasn't the only person who couldn't make it. Jae-young had returned to Seoul to find a job and Andy had dropped a computer monitor on his toe and broke both.

"Where's Tina?" I asked, looking around the room.

"She be here. She not come, but she hear you come so she drive here now."

Everyone in the family has a reputation. Yoon-chong's the artistic one, my aunt is the dry cleaner, my brother is the fat one, my father is the chemist and the one who moved to the States first, my mother is the loud one and also the one who had cancer, and I'm the one who lives in New York and apparently has the largest breasts in the world. Tina is considered the sweetest member of the family. We are opposites in personality, but she's pleasant and innocuous and speaks English and Korean fluently.

"ANNIE! OH, OUR ANNIE HAS ARRIVED!"

I whipped my head around. It was Tina and Andy's mother, my father's youngest sister, also known as the spoiled one and the one that gave me peanuts for Christmas.

"Hello!" I greeted and bowed to her. "Happy New Year!"

"You look different."

"No, I'm the same."

"You sure? You look . . . bigger." My aunt looked at my mother for confirmation.

"No, I'm the same, Aunt. You're the one who looks different."

There was something about my aunt that didn't look right. Through her blue-tinted glasses, I could see something unusual about her eyes, the shape maybe.

"She got plastic surgery," my mother whispered quickly to me. "Say thank you."

"Thank you for the peanuts," I said in Korean, "They are very . . . they are . . ." my mother nudged me ". . . there're so many of them. . . ." another nudge. "You are too generous. I love peanuts, they're so . . ." nudge "healthy."

My aunt smiled and reached over to tousle my hair but stopped in mid-air. She let out a squawk.

"What happened to your hair?"

"Nothing."

"She got it cut. It's too short, doesn't it look horrible?" my mother said, exasperated. I glared at her. Woman, I thought, you're supposed to be on my side.

"It'll grow out, don't worry," my aunt consoled me.

"I *like* my hair."

"Shh, it's OK, it'll grow out."

My father's oldest brother is in his sixties, but he seems much older. He sat in a chair and waved to me and grinned. He has a lot of gold caps on his teeth, and they gleamed under the unflattering halogen lighting. I bowed. He nodded and returned his attention to the TV. A Korean bank heist movie.

My father's youngest brother, the host of the party, approached me but stopped four feet away. He stared at me, his shiny black eyes looking me over slowly. I greeted and bowed to him. He didn't even blink.

"Is that Annie? Our Annie. Look at our Annie," he said to no one in particular.

"Happy New Year, Uncle."

"How's New York? Cold?"

"Yes. Very cold."

"Snow?"

"Not yet, but it's coming."

"You look . . ."

I crossed my arms over my chest. "I'm the same."

I heard the voices of my cousins in the kitchen and I managed to escape the living room. I always thought Yoon-chong—the artistic one—was the coolest in the entire family. When we were growing up, she drew caricatures of our family members. Naturally, my favorite was the one of my mother, done in colored pencil. She is wearing one of her favorite black dresses and her chest, rear, and hips are exaggerated. One of her hands is placed at her waist, and the other is waving an index finger in the air, as if she's scolding someone. Her mouth is opened extra-wide, with spit flying out, and her eyes are bulging red in anger. To this day, I've never shown the picture to my mother; Yoon-chong had made me promise. Yoon-chong's little sister Yoonmi is a former classically trained dancer. When she talks, she sounds just like a little girl. Her voice always rises at the end of sentences, making her statements sound more like questions. She got married a few years ago and has a three-year-old daughter whose American name is Stella. She and her husband

had wanted to name her Soma because they wanted a name associ-
ated with the heavens, but I quickly squashed that idea. "Soma," I
explained, "is a sleeping pill." I suggested Stella instead. Woo-jay,
the youngest in their family, is a mystery to me. We have very
little in common, even though we're three years apart. From what I
understand, Woo-jay likes cars and girls and that's pretty much it.

The kids sat in the kitchen, away from the adults, and enter-
tained little Stella while we waited for dinner. She is adorable, with
gigantic pigtails and a winning smile. Several relatives told me that
Stella reminded them of me when I was her age. Like Stella, I was
chatty and affectionate, and like Stella, I was spoiled. She is cur-
rently the only grandchild in the entire Choi family and for twenty-
five years, I was the youngest. We both get a lot of attention.

When Tina arrived, I gave her a hug and she sat next to me, as
she always does for New Year's dinner. Over the gigantic Korean
feast, my cousins and I stuck to idle chitchat, talking about the
weather and our jobs at the most superficial level, often combin-
ing both Korean and English into sentences so the other person
could understand. Each year, their English gets a little better, but
my Korean stays the same. Tina stepped in with translations from
time to time. My aunt passed me a plate of vegetarian wontons she
prepared just for me. I thought of Mike and wondered if he'd eat
vegetarian wontons. He doesn't consider vegetables food.

My aunt poured everyone a few tablespoons of the finest
kosher wine this side of the Dead Sea and we raised our glasses
to the New Year. The hosts wished us all good luck, prosperity,
and some other stuff I didn't quite understand. I drained my glass
and longed for more. From my spot in the kitchen, my eyes settled
on my mother's full wine glass on the dining room table. Then,
something else caught my eye. I noticed my father slowly unbuckle
his belt and unbutton his pants. They were too tight and his belly

was quickly filling up with Korean food. I nearly had a heart attack. I imagined a horrific scene where he stood up and his pants fell down to reveal his underwear with built-in ventilation, all at the dinner table. A preemptive strike was necessary. I walked over to him and stared at him hard. Then I stared at his pants. He was busted. He grinned sheepishly.

"Don't even think about it. Keep your pants on," I whispered.

During dinner my father had everyone gather around. I raised my eyebrows at my mother. She shrugged. She didn't know. My father can be spontaneous.

"I have a riddle for everyone," he announced.

Half the relatives groaned. The other half perked up at the challenge.

"There once was a man, and he wanted to go to the store. So he did. He went to the corner store. You know, the corner store? A little store? They have candy and chocolate. Cigarettes, stuff like that, not a lot of stuff. It's not like a grocery store or a supermarket. Not like in America. This was in Seoul, near where we all—well, most of us—grew up. And there was another man in the store. He was behind the counter. An old man, sitting there, or maybe he was standing. And the first man he wanted to buy some cigarettes. Maybe some milk. Or juice. He was thirsty because he was working all day in the fields. He's a farmer. The farmer wanted some cigarettes, maybe some Marlboros . . ."

I looked around and saw the faces of my relatives glaze over with a brutal combination of Korean food coma and the longest preamble to a riddle in modern history.

". . . and the cigarettes cost . . . how much are cigarettes now, one thousand won?"

Woo-jay, the smoker, shook his head. "Two thousand won."

"Wow, two thousand? I remember when they were cheaper. They were practically free and everyone smoked. People still smoke anyway, but two thousand won? That's too much. Woo-jay, you should really stop smoking . . ."

"The riddle, get to the riddle," someone said.

"So the man, the farmer, he wants a pack of cigarettes and some milk, or juice. Maybe some orange juice. Or some fruit, maybe some persimmons. The man behind the counter, he owns the store, he bought it from another man several years ago . . ."

Everyone was in agony. Yoon-chong and Stella lost interest and left the dining room. My father continued for what seemed like days telling the riddle. From what I could tell, it was more of a math problem than a riddle. Something about getting the correct change for the farmer's purchases. I wasn't clear on the details, and neither was anyone else. Yoonmi and her husband asked questions and at one point asked my father to repeat the whole story, to which my mother shook her head in fear.

"No, no, don't make him tell it again!"

Discussion about the riddle heated up around the dining room table, and unable to keep up with the Korean, I lost interest. People talked and talked, and I listened passively. At the request of my father and an uncle, Woo-jay got out a pen and paper and started writing details down. Yoon-chong approached me, her eyes filled with pain.

"Please, please, make him stop. I'll do anything. Please, this is horrible. Why is he still talking?" she asked in Korean.

"There's nothing I can do," I answered in English.

"But he's your father. Can't you turn him off?" She pretended to push a button on a remote.

"It's not working, let me try." I took the imaginary remote from her and banged on it to get the batteries to work. She laughed hysterically.

"Is there a mute button?" she wanted to know.

"I think we all can use a mute button."

Yoon-chong nodded gravely. "You are very wise." She bowed to me and I laughed. We listened to my father repeat the riddle. I caught my mother's attention and pretended to fall asleep with my head tilting down to one side. She stifled a giggle and winked at me.

"My husband, please, is there a point?" she asked.

"Yes, there's a point. It's a riddle."

"Are you sure? It doesn't sound like a riddle."

The discussion continued. Yoonmi and her father acted out the transactions between the storeowner and the farmer using dollar bills. I started fading. Tina punched me in the arm.

"Hey, you OK?"

"Coffee, " I gasped, "I need coffee. Badly."

"We have to peel fruit first."

For the millionth time that day, I groaned. Koreans end their meals with fruit, *peeled* fruit. Years ago, my mother told me that one sign that a Korean woman will make a good wife is how well she can peel fruit. The other signs include expert sewing and having a long second toe. (People have also told me that long second toes are signs of aristocracy, prosperity, intelligence, good luck, laziness, and a short life span. I have very long second toes.) Tina brought over gigantic bowls of Asian pears, apples, and persimmons.

"Oh God, do we have to peel *all of that*?"

Yoon-chong held Stella on her lap and was excused from peeling. Tina began peeling and cutting the fruit in perfect little pieces. She peeled the skin off the apples in one long strip that curled around, which fascinated both Stella and myself. She handed me a knife.

"I can't peel fruit. I suck at it. Honestly. Why don't they just eat it with the peel? That's where all the vitamins are anyway."

Tina laughed and translated for Yoon-chong. My hands awkwardly maneuvered over an apple and by the time I finished peeling, it had turned brown. I was hopeless.

"Three hundred and sixty three won!" Woo-jay whooped and clapped his hands, "I got it!"

"Wrong!" my father cried.

Everyone groaned. Aunts began clearing plates and washing dishes. Eventually the riddle was resolved and Yoonmi came into the kitchen to help us peel fruit. She grinned triumphantly.

"I won ten dollars for getting it right."

The kitchen was bustling with ladies, old and young, and the men moved into the living room to do what they do best—nothing. At least that is what their wives tell each other.

We brought fruit out to the living room where everyone gathered for the traditional bows. Every New Year's Day, children bow to their elders and wish them good luck and prosperity. The elders dispense advice and then present cash. Sometimes, a lot of it, enough for a nice dinner and a night of irresponsible drinking.

"Now that we have a grandchild in the family," my uncle announced, "we will not be giving out money."

"WHAT?"

My jaw dropped. I'm not going to lie, I like money. I like what it can do for me. The kids had a good scam going—bowing for bucks.

"Why? I don't see how a grandkid changes anything," I protested in my shaky Korean, "if anything, we should get more money."

My relatives laughed.

"Why would you get more money?" my mother asked me. Again, she was taking the wrong side.

"I don't know, but really, we should get some money. It's tradition."

I looked around for support from my cousins. Yoon-chong grinned in approval. I motioned for her to help me out. She shook her head in fear. The cause would have to be my own.

"Everyone's too old to get money," my father retorted.

"What are you talking about? You're never too old for money."

Everyone laughed except my mother, who rolled her eyes. Her daughter, she thought, was being too sassy.

"Our little Annie is smart," my uncle said, "but she will remain poor. No money, sorry."

My cousins thanked me for my effort. I shrugged. I fought the good fight.

Normally, the kids bow with their siblings in front of each pair of adults. Since Tina and I were without our brothers, we decided to bow together. This only emphasized how clumsy my bow looked next to hers. In addition to being known as the one in New York and the youngest one, I'm also known as the worst bower.

"Annie, are your legs broken?"

"It's like her legs just disappear. Pulverized completely."

"Everything from the waist down just vanishes all the sudden and boom, she's on the floor."

"Is she bowing or falling? I don't understand."

This year's advice to all the young girls was marriage. Yoonmi ruined the rest of us by getting married and having a kid. Every pair of adults kept asking us when we were going to get married. Yoon-chong, being in her thirties, simmered. She had to answer this question on a daily basis.

"When are you going to get married? When? How long must we wait?" my aunt wailed melodramatically.

"Right after I get a boyfriend," Yoon-chong replied.

"When I meet the right person," Tina replied.

"Never," I said. "Not with you people in the family."

"ANNE!"

"I'm joking, I'm joking."

My aunts and uncles laughed and my mother apologized for my behavior. "I don't know where she gets her mouth."

Three-year-old Stella bowed gracefully in front of her grandparents. Everyone clapped for her and showered her with smooches and hugs and praise. She was so cute I wanted to stick her in my pocket.

"Annie, are you watching this? You should bow like Stella. Maybe she can teach you how," my uncle chided.

"If you give me money, I'll bow better."

"ANNE!"

When Stella bowed in front of my parents, my father reached for his wallet.

"No, no, no, don't give her money, she doesn't understand it," Yoonmi cried. She waved her arms to stop my father.

"We don't want to teach her about money yet," Yoonmi's husband said respectfully. He has a deep voice that can fill any room. He explained that they wanted to wait until Stella could understand its value and danger.

My father brushed them aside and started sifting through his wallet. Yoonmi and her husband exchanged furtive glances.

"Dad, they don't want you to give her money. So don't. Stop."

He counted out five one-dollar bills for Stella. I shot him the evil eye, which he ignored. She triumphantly held up the money, not knowing exactly what it was, other than the fact that it was good. Her parents remained stiff. The rest of the aunts and uncles followed in suit and doled out all their singles. Stella grabbed a fistful of dollars and squealed. My aunts and uncles asked what she wanted to buy and she started naming off dolls and toys. She wanted to go to Toys "R" Us immediately and buy everything. In under an hour, the adults unleashed the consumer inside Stella. Yoonmi threw up her hands.

"Sorry," I whispered, "my dad started it all."

Suddenly I smelled something glorious from the kitchen. A waft of heaven floated into my nostrils and I felt, for the first time that day, completely relaxed. Coffee. Regular coffee. Real coffee. Not instant. Not with vanilla or hazelnut or fudge or mint raspberry pumpkin spice. Just coffee. Simple, black, mine. My aunt poured me a cup, and I took a few euphoric sips, feeling caffeine slowly trickle through my veins. I stood next to the fireplace silently, with my eyes closed, enjoying a brief moment of bliss, just Juan Valdez and me. Yoon-chong's hand on my elbow woke me up. She asked me to help prepare the *yut* board. I set my coffee down on the mantle.

The traditional Korean board game of yut is played on a home-made board. I think back in the day the board was drawn in the dirt. Every board is a little different, and the one at my uncle's house is drawn with colorful marker on poster board. The spaces form a square and there are two diagonal lines that run across the middle to connect the four corners—a "shortcut." The board looks like the one used in Sorry! Each player or team has three or four markers, and the goal is to get all the markers to the finish line, either by going around the board or cutting across the middle. Instead of dice to determine the number of spaces each marker can advance, yut uses four sticks. The sticks have Chinese characters on one side, but I don't know what they mean. A player throws all four sticks into the air and the way they land determines the number of spaces a marker can move. If one stick lands face-down, the player can move one space. This is called *do* (pig). If two sticks land face-down, the player can move two spaces. This is called *gae* (dog). If three sticks land face-down, the player can move three spaces. This is called *geol* (chicken). If all the sticks land face-down, the player can move four spaces. This is called *yut* (cow). If all sticks land face-up, the player can move a marathon of five spaces. This is called *mo* (horse). Throwing a mo also allows the player to toss the sticks again. I think the animals are ranked according to speed. The horse is the fastest, so it gets the most spaces attributed to it. But I've always thought a dog was faster than a chicken, so maybe I'm wrong.

There are a few twists to the game. If a player's marker lands on a space occupied by an opponent's marker, the opponent has to start over from the beginning of the board, and the player gets to throw again, just like in Sorry! If a player's marker lands on a space occupied by their own marker, the markers become connected and move as one piece. This is dangerous because if someone else lands on you, you'll get screwed. On my uncle's board, there's a special space that allows the marker that lands on it to move directly to the finish line.

After I helped my cousin set up the board on the floor, I went back to the mantle to retrieve my coffee, only to find it was gone. My aunt had cleared my cup in the five minutes I was gone. She was fluttering around the dining room and living room furiously cleaning up after guests. I poured myself another cup, took a few sips, and set it down on the coffee table. The teams were divided by family and each paid five dollars to the pot. The winning family would receive fifteen dollars. I suggested mixing up the teams—I wanted to play with Tina or Stella, not with my mother. She's a fiercely competitive person.

"You ready, Anne? You ready for win?"

My mother threw a few warm-up tosses with the sticks. I could see the aggression in her eyes; she was in it to win it. My father watched my mother and laughed. The men in the family, though technically playing the game, sat on the couches and chatted amongst themselves. The rest of us gathered on the floor around the yut board. Tina threw first for her team and got a mo. She advanced her marker five spaces and got to toss again—a terrific way to open the game. Her mother cheered a little too loudly.

"Tina's going to win this for us! Right Tina?"

They high-fived enthusiastically and passed the sticks onto Yoon-chong's family. Stella threw for her team first and got a geol, advancing their marker three spaces. Everyone cheered; this was her first yut game. The oldest aunt, the wife of my father's old-est brother, tossed the sticks and advanced on the board quietly. All of her children were in Seoul, so she played alone with little fanfare. When the sticks came around to us, my mother looked at me intensely. The game had barely begun and she was already feverish.

"OK Anne. You throw big. Get mo like Tina! Get big number so we can win! GO!"

I tossed the sticks and got a do, one space.

"Agh, *Anne,* why only *do*?"

"Mom, relax. We just started."

"But you not throw good. You have throw better. Use arm, like this."

She demonstrated with her arm, curling it forward, then thrust-ing her wrist, and finally releasing her fingers.

"Throwing has nothing to do with it. It's luck."

"You make Mommy lose."

"What are you talking about? We just started the game!"

I looked at the board and hoped that whoever beat us would do it swiftly. I reached for my cup of coffee, only to find it was gone again. I grunted out of frustration. My anal-compulsive aunt was going over the line. I got up for another coffee.

"Anne, where you go? You have play game!"

"To get coffee, can you just chill out, please?"

I poured myself a third cup and tried to guzzle it, but it was too hot. Determined to finish it, I hid the cup behind the poker for the fireplace. Tina's mother was up. She held the sticks tightly and flung them to the floor. They advanced a new marker onto the board. When it became our team's turn to toss again, my mother picked up the sticks. She tossed a yut and landed on Tina's marker, sending it back to the starting line. My aunt yelped loudly like a wounded animal, and Tina consoled her. She explained that they were still in good shape. My mother grinned triumphantly and clapped.

"That how you throw, Anne. See how good Mommy play?"

"But it's based on *luck*."

"No Anne, it skill."

"Skill? There's no skill involved here. Zero. It's like dice."

I looked at Tina for help and she smiled sympathetically. If my brother were here, he would've agreed with me.

The game continued and Yoon-chong's marker landed in a space occupied by ours, and their team erupted in cheers and high-fives. Our marker was sent to the starting line. My mother became irate.

"That wasn't a gae, that was a geol! See? One of the sticks is on its side. So it's three spaces, not two. Leave our marker on the board," she sputtered in Korean.

"No it's a gae," my oldest aunt said.

"If it's on the side, it counts as face-up. It's gae," Tina's mother confirmed.

"Definitely gae," said an uncle.

My mother stewed at the injustice. "Mom, it's just a game."

"Anne, don't you want win?"

"No, not really."

"Don't you want money?"

"Sure, but it's just fifteen dollars. The only thing you can get with that is coffee."

Coffee! I looked for my cup behind the fireplace poker. Missing. I sighed. Why didn't I just guzzle it when I had the chance? I got up to pour myself another cup of coffee.

"I think you drink too much coffee. Not good for you."

I slipped into the kitchen only to find the large percolator empty and my anal-compulsive aunt scrubbing away. She'd rather clean than play yut with a bunch of animals.

"How many cups have you had? No wonder you're so short. There's no more coffee, sorry," she apologized in Korean. "You can make more."

"Really?"

She handed me a family-sized jug of Folgers crystals. Instant. The container was so large, it had a handle. I was desperate, but *instant*? To me instant coffee is when you order it and someone hands it to you. My mother called me from the living room. It was our turn again.

My father, who got intrigued by all the commotion and joined the game, threw the sticks and moved our marker on to the same spot as another one of our markers, allowing both markers to travel together. My mother beamed.

"Anne, look how you Daddy play yut, he so good!"

"No skill, Mom. It takes none at all. Come on, Dad. You know this; do the math."

I tried to appeal to my father's scientific mind. In his brain floated chemical formulas and ionic bonds and the Periodic Table of Elements. He had to agree that four sticks tossed randomly would produce a random result.

"I think it take luck, but luck take skill."

"That doesn't even make sense."

Tina's team was in the lead. They had gotten three of their four markers to the finish line. Both Yoon-chong's team and my oldest aunt were tied for second; they each had two markers on the finish line. We were last, with only one marker on the finish line.

"Anne, we have to win! Get mo! Get five space! Throw like Mommy, like I show you!"

I tossed the sticks, exaggerating the way my mother threw. I curled my arm and swung it back deeply, as if I was bowling. I lengthened my arm and hurled the sticks onto the floor and followed through with my entire arm, raising it straight into the air. My relatives burst into laughter. Even my mother grinned.

"How about that?"

"Anne, you such clown. Look what happen!"

Our marker landed on the same spot as the other two markers. We had all three on one space; all our eggs were in one basket. Our relatives smelled weakness. Tina's mother shouted to her daughter to toss a number that would land on our space. Yoon-chong and Yoonmi cheered for Stella to knock us out. Even my oldest aunt, the soft-spoken one, held the sticks closely to her heart and prayed for a number that would send our team back to the starting line. They were out for blood. My mother was not amused. She shook her head.

"Everyone so crazy here! How come we not win?"

"It's luck. I have bad luck, that's all. That's it."

"Why you have bad luck?"

"Why does anyone have bad luck? It's just luck."

"Maybe you practice more."

"Practice *what?*"

Finally the last round of tossing began. Tina's final marker was close to the finish line. They would win the fifteen-dollar pot. I picked up the sticks and threw them listlessly, with the same sense of surrender as a team hopelessly down with three seconds left in the game.

"Oh my God! *Anne! Oh my God!*"

"What?"

"WE WIN! WE WIN!"

I had landed all three of our markers on the special space that sent us to the finish line before Tina's team. My mother pumped her fists into the air and clapped loudly. She grabbed my shoulders and shook me excitedly. It was as if we had won a $15 million jackpot. Tina's mother was shocked.

"No, what is this? That's not how we play. That doesn't count. Who put that space on the board? That's not fair. It doesn't count!" She tried to yell over my mother.

"Anne, you know song—'We are champion!'—you know? We have to sing song."

I laughed, imagining my mother rocking out with Freddie Mercury.

I grinned. "Mom, I'm so not singing that song."

My aunt looked at the yut board and shook her head. Tina squeezed my shoulder.

"See? You're not bad at this game."

"It takes no skill, none at all."

My mother motioned to high-five me, something I don't do. My New Year's resolution back in 2003 was to be high-five-free. It started as a joke, but I actually did feel silly and uncomfortable when I high-fived, so I just cut it out of my life. Among my friends, I'm known as the one who doesn't high-five.

"Mom, you know I don't high-five."

I left her hanging but she laughed and hugged me anyway. She shoved the fifteen dollars into my pocket.

"Buy Mommy something nice!"

After a while, the conversations in the living room slowed down and relatives lowered their voices because Stella had fallen asleep on the couch. Close to midnight, everyone stood up to collect their coats and bags. As I put on my shoes to leave, I saw a hand stuff something into my bag. I looked up; it was my anal-compulsive aunt.

"I know we're not doing money this year," she whispered, "but your young, you live in New York. You need it."

"No, no, please, I was only kidding before."

She hugged me and handed me a bag of neatly wrapped vegetarian Korean food. I thanked her and wished her a happy New Year again and stepped outside. The fresh air felt crisp and it smelled like damp grass, something I rarely smelled in New York. I realized that for the past several hours I had been breathing garlic, sesame oil, and a heavy combination of musky and flowery fragrances (my father wears Drakkar Noir). I bowed to my aunts and uncles and said good-bye to my cousins as they each got in their car. I felt like it was the end of a movie, when each person walks off into the night, in a different direction, and the camera pulls back from a peaceful landscape while the credits roll. Tina's mother took me aside.

"You are still the youngest in your generation. You're our Annie. I'm the youngest too, you know."

"I know."

"Work hard."

"I always do."

She pressed some bills into my hand and floated away. I fished the keys out of my bag as I walked out to my car. My parents came up beside me.

"Anne, you know how to use this? Someone gave me as gift and I don't know how to use. You should take it." He held out an American Express Traveler's Cheque. He winked.

"Thanks."

"I see you at home later. Don't stay out too late with your friends."

With a wave, my father walked off to his car.

"This jacket so thin, you need new one. You not cold?" It was a cold evening by Los Angeles standards and my mother buttoned up my favorite jacket—the cuffs are worn soft and thin and she always suggests it's time to let go.

"No, this is nothing compared to New York."

She carefully closed my car door, looking to make sure my fingers were out of the way—something she's done since I was seven when Mike accidentally slammed the car door on my pinkie at Dodger Stadium. I turned the key and the engine stirred gently. I rolled the window down.

"Good night, I'll see you later!"

"Happy New Year, Anne. You live! You survive!"

As I drove off, I stuck my arm out the window and waved. I watched my mother in my rearview mirror as she waved back and smiled.

ACKNOWLEDGMENTS

For the most part, writing makes me want to gouge my eyes out with a fork, so I deeply appreciate the people who have removed all the sharp objects within my reach. I was lucky to have two editors who worked tirelessly to wrangle with my words. Jill Schwartzman asked all the right questions and kept me on the path of awesome freshness. Jeanette Perez remained organized and calm and tightened every last screw in the project. My agent, Douglas Stewart, supported my every step and assured me that "this doesn't suck as much as you think."

My professors at Columbia University pushed and prodded, and I am grateful I had the opportunity to work with them. Many of the stories in this book were crafted under the careful guidance of Richard Locke and his magical beast of a brain. Patricia O'Toole is the greatest cheerleader any writer could hope for; she must be cloned and available to all. Vince Passaro told me what was funny and what wasn't. The wily Englishman, Michael Scammell, got me started and fed me warm vittles that thankfully were not English.

Aura Davies, Sarah Smarsh, and Rhena Tantisunthorn rallied to my aid with comments, line edits, coffee, and beer. And more beer. And then more coffee. Aaron Isaksen has supported me since the very beginning, back in the days of yore. Sanford Kaye at Harvard Extension School, Nathan Bowers, Michah Calabrese, Cristine Gonzalez, Landon Hall, Chris Leong, Mika Oshima, Dave Schaye, Rosalyne Shieh, and Michele J. Thomas cut, blow-dried, and styled much of my writing and offered design advice. Perri Pivovar took my photos and suffered mosquito bites and missing bike messengers in the process. I am lucky to be in the company of such creative minds and generous friends.

Finally, I thank my parents for being irritating enough to warrant a book and loving enough (I hope) to forgive me for writing it. I also thank my brother, Mike, who has a warm and gentle heart even though he tries to hide it from me. I truly appreciate the support of my large, disruptive extended family, especially my cousin Andy. They know all too well the joys and horrors of growing up with me. Mostly joys.